SpringerBriefs in Ethics

For further volumes:
http://www.springer.com/series/10184

Daniel K. Sokol

Doing Clinical Ethics

A Hands-on Guide for Clinicians and Others

 Springer

Daniel K. Sokol
Department of Primary Care and Public Health
Imperial College London
Charing Cross Campus, Reynolds Building
St Dunstan's Road
London W6 8RP, UK
e-mail: daniel.sokol@talk21.com
URL: www.medicalethicist.net

ISSN 2211-8101   e-ISSN 2211-811X
ISBN 978-94-007-2782-3   e-ISBN 978-94-007-2783-0
DOI 10.1007/978-94-007-2783-0
Springer Dordrecht Heidelberg London New York

Library of Congress Control Number: 2011941665

Printed on acid-free paper

Springer is part of Springer Science+Business Media (www.springer.com)

...an admirably short and clear guide to doing medical ethics... I welcome this book and urge medical students and doctors of all grades to read it in paper, on-line, or on their portable screen reader.

From the Foreword by Sir Richard Thompson, President of the Royal College of Physicians, UK

Dr. Sokol has provided the field with a much needed, easy and comprehensive tool on 'doing' clinical ethics that all should have in their back pockets.

Dr. Nneka Mokwunye, Director of Bioethics, Washington Hospital Center, Washington DC, USA

This is a magnificent guide to clinical ethics and reflects the author's very well known and widely respected academic gravitas and real life experience in clinical ethics. It is a must read for anyone involved in the field.

Mr. Vassilios Papalois, Consultant Surgeon and Chairman, Imperial College Healthcare NHS Trust Clinical Ethics Committee, UK

For Sam

Foreword

Dr. Sokol is a senior lecturer in medical ethics who, for some years, has studied and taught the subject at Imperial College London and other institutions. As a result, he has written an admirably short and clear guide to *doing* medical ethics, aimed at medical students and practising clinicians. In this guide, the difficult but fundamental vocabulary of beneficence and maleficence, etc. is explained (or, in my mind, doing good and doing harm) in an effort to provide quick but reasoned answers at the coal face.

Many clinicians are turned off by ethical problems, probably because they are often much less clear-cut than those of a clinical nature. As Dr. Sokol says, there are often several right answers to an ethical problem, which is why he emphasises that this is a personal guide. Clinicians, on the other hand, are taught necessarily to decide quickly on one course of action that, at a given moment, seems to be in the best interests of the patient.

Perhaps some of us are also put off by those serious, even deep, discussions, on and off the media, of clinical examples. These are usually discussed by thoughtful (can one be too thoughtful?) and probably highly intelligent ethicists, who seem to make difficult decisions more difficult, and soon slip into philosophy. This sits uneasily with rapid clinical decision making. Many hospitals have a standing committee available to help resolve less urgent problems but, when decisions are not straightforward, most of our advice is obtained from experienced nurses and colleagues, and from families and carers.

This book is engagingly written, devoid of abstruse philosophy, and rich in practical, down-to-earth advice. There are also useful chapters on writing about medical ethics, teaching ethics, and asking for ethical permission to carry out clinical research, topics that are not usually found in textbooks.

I welcome this book and urge medical students and doctors of all grades to read it in paper, on-line, or on their portable screen reader. Dr. Sokol talks about one's ethical brain, or, as I see it, an ethical elf always sitting on one's shoulder and watching. It can be trained by considering problems and discussing them with the elves of friends and colleagues and, of course, by a careful reading of this book.

My only regret is that he did not digress into the ethical controversy of the day, namely assisted dying, but I hope that will be for another book!

London, August 2011   Sir Richard Thompson
                                                          President, Royal College of Physicians

Acknowledgments

I wrote most of this book in Aix-en-Provence, in the south of France. My mother, a chef to rival Paul Bocuse, sustained me with sumptuous meals during the long days at the keyboard. My father, as well as reviewing the manuscript, dragged me onto the tennis court to prevent muscle atrophy from prolonged sitting. And Sam, my ever-patient wife, adopted a laid-back *Provençal* attitude, and cheerfully left me to my solitary work. To them all, I am grateful.

On my return from France, a rough draft in hand, I asked friends—perhaps now former friends—to comment on the draft. I am indebted to Henry Mance, an eagle-eyed journalist, for his characteristically honest suggestions, to Dr. Philip Sedgwick, statistician extraordinaire, for reviewing Chap. 4, to Richard Warry, editor at the BBC News Online, for his tips on submitting pieces to the BBC, and to Alan Cedrik, journalist and chess master, for his helpful critique of an early draft.

I am also grateful to my learned friends, Aidan O'Brien, Mathew Roper, Tom Bradfield, Susan Jones, and Leo Meredith, who displayed the kind of punctiliousness characteristic of drafting experts. One long and memorable comment on the proper order of a question mark, a quotation mark, and a full stop ended dramatically "you have effectively fused two sentences, leading to a punctuation quandary".You know who you are.

Countless thanks are also due to the clinicians and ethicists who took time from their busy schedules to read the book. Their suggestions have been invaluable. They are Drs. Zuzana Deans, Fauzia Paize, Hannah Peters, Francesca Rubulotta, Alifa Isaacs-Itua, Lynnette Hykin, Sabaretnam Muhundhakumar, David Hunter, Ayesha Ahmad, and James Wilson.

A special note of gratitude goes to Raanan Gillon, Emeritus Professor of Medical Ethics and erstwhile supervisor of my PhD, whose influence permeates much of this book, and to Sir Richard Thompson, President of the Royal College of Physicians, who generously agreed to write the Foreword.

Thanks also to Tai Sae-Eung, a talented young artist from Boston, USA, who provided the original artwork for the book, and to Meagan Curtis, from Springer, who made the whole publishing process so smooth.

Finally, this book would not exist were it not for the hundreds of medical students, patients, and colleagues who have taught me many of the lessons contained within. Thank you.

All errors and omissions that remain in the text are, of course, mine alone.

Contents

About the Author

Daniel Sokol is an Honorary Senior Lecturer in Medical Ethics at Imperial College London, and a barrister. He read Linguistics and French at St Edmund Hall, Oxford University, and completed Master's degrees in Medical History and Medical Ethics, and a PhD in Medical Ethics at Imperial College. He has held academic positions at Keele University, St George's University of London, and Imperial College, and has taught medical ethics to students and clinicians at various institutions across the world. He has been a visiting scholar at hospitals in the United States, Canada, and India, where he has worked closely with hospital ethicists and clinicians. He has published widely in academic journals, newspapers, and online. In 2005, his first book, Medical Ethics and Law, co-authored with Dr. Gillian Bergson, was runner-up in the Book of the Year Award of the Medical Journalists' Association. Since 2007, he has written the 'Ethics Man' column for the British Medical Journal. That same year, he founded and directed the UK's first course on applied clinical ethics for doctors, nurses and members of clinical ethics committees. It continues to run under his co-direction. He is a Senior Editor of the Postgraduate Medical Journal, commissioning and reviewing articles on medical ethics and law. He sits on a number of committees, including those of the Royal College of Surgeons of England, the Ministry of Defence and the Ministry of Justice. He lives in London, England.

Introduction

This book is a personal view. It is one ethicist's opinion on how to do ethics in medicine, developed over years of trial and error. It does not claim to be *the* right way. There are countless ways to skin this particular cat, but this is the way I do it. It has worked for me. Some of the errors, many committed by me and some by others, are recounted in these pages so that readers may learn from them.

This book is not an academic text. It is strewn with anecdotes, and is written in simple, almost conversational, language. In my quest for effective communication, I have taken the liberty of addressing you directly, much as I would if advising a friend in the Distiller's Arms down the road. The anecdotes are included as an aid to learning, an antidote to boredom, and as supporting evidence.

This book is not the place for an overview of the ever-growing bioethics literature, or for stimulating debate about the nature of autonomy or the source of moral norms. The brevity of the work demands a direct approach. It is primarily a practical guide rather than a textbook, and as a practitioner myself I know that when consulting a book of this type I want 'answers' quickly, without trawling through academic debate.

Although there are thousands of books and articles on medical ethics, there is little written exclusively on the subject of actually *doing* clinical ethics. This is the ambitious aim of this slim volume: to teach clinicians how to apply ethics at the coal face. It contains professional secrets that, a few years ago, I would have been reluctant to share. With age, my competitive spirit has waned, and so this book is a crib sheet of advice to the aspiring ethicist[1].

---

[1] Benjamin Cardozo, a celebrated judge of the US Supreme Court, commented on the secrecy of judicial decision-making: 'Any judge, one might suppose, would find it easy to describe the process which he had followed a thousand times and more. Nothing could be farther from the truth. Let some intelligent layman ask him to explain: he will not go very far before taking refuge in the excuse that the language of craftsmen is unintelligible to those untutored in the craft (Cardozo 1921).' More recently, the barrister David Pannick wrote: 'Like members of the Magic Circle who face expulsion if they explain how the trick is done, judges are eager to protect the mysteries of their craft (Pannick 1987, p. 10).' Andrew Soltis, a chess grandmaster, has also observed that

The book provides analytic tools to help identify and resolve ethical problems. It also describes the language of ethics, through which your views and reasoning about those problems may be expressed more clearly and forcefully. The tools are valuable both to construct your own arguments and to attack those of others.

The great physician William Osler once remarked "He who studies medicine without books sails an uncharted sea, whereas he who studies medicine without patients does not go to sea at all" (Bean and Bean 1950, p. 28). This book will only get you so far; ethical skills are honed by application to real cases. Just as a criminal lawyer yearns for cases involving murder, rape, and other grave offences, so too should clinicians embrace difficult ethical problems as opportunities to develop as ethicists.

The book is entitled 'doing medical ethics'. The phrase is interpreted broadly to capture the application of ethical knowledge to a concrete situation in the field of medicine. Doing ethics goes beyond resolving ethical problems in clinical practice. Writing an article on clinical ethics is doing ethics. So too is a presentation on ethics at a conference, or teaching others about medical ethics, or trying to get your research approved by a research ethics committee.

The book is divided into four chapters, each largely self-contained. The first is on ethics in the clinical environment. It is, by far, the longest chapter, reflecting the richness and complexity of the topic. The second is on publishing and presenting in the field of clinical ethics. The third focuses on teaching clinical ethics, and the final chapter is on applying for research ethics approval. Chapters 2–4, by virtue of their highly practical focus, are written in a more prescriptive tone. I tell it as it is, or at least as I see it.

I have no doubt that my direct approach will raise eyebrows among colleagues of a purer bent, for whom the work will seem simplistic. I can only remind readers of the introductory nature of this text, and re-affirm that the methods and advice contained within have served me well.

References

Bean R, Bean W (1950) William Osler: aphorisms from his bedside teachings and writing, Henry Schuman Inc., New York

Cardozo B (1921) The nature of the judicial process: the method of philosophy, Yale University Press, New Haven. http://www.constitution.org/cmt/cardozo/jud_proc.htm. Accessed 3 August 2011

Pannick D (1987) Judges, Oxford University Press, Oxford

Soltis A (2005) How to choose a chess move, Batsford, London

(Footnote 1 continued)

'Traditionally [chess] masters have revealed little about how they actually choose moves (Soltis 2005, p. 5).' This reluctance to expose in plain view the exact process of decision-making is, in my opinion, also found among professional clinical ethicists.

Chapter 1
Clinical Ethics at the Coal Face

There are clinicians filled with a passion for medical ethics, and there are those who could not give a fig for it. Yet, love it or hate it, there is no escaping the fact that, for most clinicians and medical students, ethics is part of the job, just like paper-work. Since it cannot be avoided, any self-respecting clinician has reason to learn to 'do ethics' and should strive to do it well.

There is another reason for clinicians to learn this skill. The American judge Oliver Wendell Holmes Jr. wrote in *The Path of the Law* about the importance of 'getting the dragon out of its cave' (Holmes 1896). The dragon represents the rule of law underpinning a legal decision. "When you get the dragon out of his cave on to the plain and into the daylight", Holmes continues, "you can count his teeth and claws, and see just what is his strength" (Holmes 1896, p. 20). This was a call for judges to provide openly the reasons for their judicial opinions, so that these could be subject to public scrutiny. The message applies equally to ethical decision-making in medicine, where the stakes, as in law, can be of enormous significance to the welfare of individuals.

D. K. Sokol, *Doing Clinical Ethics*, SpringerBriefs in Ethics 1,
DOI: 10.1007/978-94-007-2783-0_1, © The Author(s) 2012

Clinicians who make important ethical decisions should be in a position to expose the rationale for their view. This ability should form part of the skill set of any reasonably competent clinician, and falls broadly under the principle of beneficence, or acting in the best interests of the patient. Poor reasoning can lead to bad decisions, and consequent harm to the patient and others. Knowing how to do ethics is, therefore, a professional and moral obligation.

The process of ethical decision-making can be divided into three stages (Andre 2002, p. 78):

1. Moral perception (seeing the ethical problem)
2. Moral reasoning (resolving the problem analytically)
3. Moral action (implementing the chosen solution)

1.1 Step 1: Moral Perception

A friend points to a cloud. You look up and see nothing but a cloud. He then says that the cloud looks like a face, and suddenly you recognise the features of the face: the nose, the mouth, the eyes.[1] Moral perception describes a similar effect in

---

[1] For reasons of convenience and stylistic fluency, the masculine form for pronouns is habitually used throughout the book. As lawyers say, 'unless the contrary intention appears, words importing the masculine gender include the feminine' (Interpretation Act 1978, s6(a)).

the context of ethics. It refers to an awareness of the morally salient features of a situation; features that may not, at first sight, be obvious. It is important because, in real life, ethical issues do not come pre-labelled. No finger points helpfully to the problem. Without moral perception, ethical issues float past never to be resolved.

I used to present a scenario to medical students and ask them to identify any ethical issues. "A registrar on your firm asks you to do lumbar puncture on a patient. The patient is a middle-aged Afro-Caribbean man who is an IV drug user. You have never done a lumbar puncture before, but you did see one last month. The registrar prepares the patient, and stares at his bleep. 'He's all yours', he says, as he rushes out of the room."[2]

Some students, despite sitting in an ethics class, would 'see' nothing or make irrelevant observations. Others would spot a constellation of ethical issues and points of discussion: the challenges of inexperience, the validity of consent, the possibility of causing harm, the perceived need to impress and be a 'team player', the inadequacy of supervision, the effect of race and perceived social worth on patient care, and even the relevance of geographical context, local resources, and staff availability.

This observation about variable moral perception is reflected in a study published in the *British Medical Journal* in 2003 (Caldicott et al. 2003). The study revealed that nearly a quarter of intimate (rectal and vaginal) examinations were performed by medical students on anaesthetised or sedated patients without the consent of those patients. The authors interviewed some of the medical students. Some did not recognise the ethically problematic nature of the examinations. "The patients will never know", they said. The moral issue was not on their radar. Others knew that it was wrong, that respect for autonomy required them to seek permission first, but they felt compelled to go ahead. Students in the same year, from the same medical school, ranged from the morally blind to the morally astute.

The failure of moral perception can arise from outright ignorance of ethical matters. In an unfamiliar area of medicine, the issues may be invisible to untrained eyes. A urologist may be ignorant of the ethical issues in ophthalmology, and vice versa. A clinician might not know that adding a colleague to a publication is ethically inappropriate if the colleague has failed to fulfil the authorship criteria.[3] Like fasciculations in a patient with motor neurone disease, they only become apparent once they are pointed out. Once identified, the issues are often obvious.

In military medical ethics, for example, a major issue concerns the decision to treat severely wounded patients from the local area. The difficulty is that, once stabilised in a state-of-the-art hospital, the patient is released into a substandard healthcare facility, where he will probably die.[4] This ethical issue can easily be missed by an ethicist unfamiliar with the practice of military medicine in conflict zones, such as Afghanistan. To be morally perceptive, then, a sound grasp of the

---

[2] This scenario is adapted from Kushner and Thomasma 2001, p. 33.

[3] See Appendix 1 and Chap. 2 for a discussion of this aspect of publication ethics.

[4] I explore some dilemmas in military medical ethics in an article reproduced in Appendix 2.

realities of the situation on the ground is necessary. So too is knowledge of the moral norms operating within it. Without knowing that respect for confidentiality is a moral norm, the fact that two nurses are talking openly about a patient in a crowded lift will not trigger your moral radar.

Another cause of poor moral perception is what can be termed the 'heat of the moment'. On a busy ward or operating theatre, the many distractions can make you less clear-sighted than usual. It is only later, in retrospect, that you wonder how you missed the moral issue. In an article in the BMJ, I described a visit to a nephrology ward round in which I was lulled by the prevalent clinical mindset (Sokol 2007):

> It [the ward round] proved to be a puzzling experience, not because the blood gases, creatinine levels, diagnostic tests, and myriad statistics recited by a junior doctor sounded like one of Mallarmé's incomprehensible poems, but because, as the afternoon progressed, I noticed the patient-as-person fading behind this shroud of science. I felt comfortable with my consultant, my team with their dangling stethoscopes, the all-knowing computer wheeled by the bedside, and the timid patient, dwarfed by our confident crowd. Ethics seemed a million miles away.[5]

On a ward round, in a morbidity and mortality meeting, a multi-disciplinary meeting, or in other situations where medical matters are at the forefront of the clinician's mind, the ethical part of the brain can lie dormant, unable to recognise even prominent moral features. The clinician operates in 'medical' mode, to the detriment of the ethical brain. Yet, the importance of moral perception is obvious: without it, ethical problems go unidentified and unresolved. And without early diagnosis and intervention, they may grow like tumours into full-blown catastrophes.

1.1.1 Improving Moral Perception

I once set an examination question on futility. At the pre-examination review meeting, a consultant admitted that he got the question wrong. When asked how he would define futility, he paused thoughtfully before declaring: "Something's futile when I say it is". The lesson is that some terms, such as 'futility' and 'best interests', are only seemingly medical. The clinician's moral values are, in fact, concealed under a cloak of objectivity.

1.1.1.1 Futility

'Futility', a word often used to justify withholding or withdrawing treatment, is a good example of a term with a hard, scientific exterior but a soft, subjective core. The more senior the clinician who uses it, the more objective it seems. It is, however, value-laden.

---

[5] A complete version of the article is available in Appendix 3.

Futility is goal specific. In other words, the futility or otherwise of an intervention depends on its goal. Cardiopulmonary resuscitation may be futile if the goal is to restore normal cognitive function, but not futile if the goal is to prolong life for a few days. The goal may even be non-clinical. On the battlefield, a combat medic may treat a gravely injured soldier primarily to maintain the morale of the troops. It is legitimate, when told that an intervention is futile, to ask "futile with respect to what?"

Jonsen et al. (2010) provide a helpful distinction between types of futility:

1. **Physiological** futility is when the intervention cannot physiologically achieve the desired effect. This is the most objective, and least controversial, type of futility. If a patient has an illness caused by gram-positive bacteria, for example, it would be physiologically futile to administer an antibiotic that is effective only against gram-negative bacteria (Lo 2000, p. 73).
2. **Quantitative** futility is when the intervention has very little chance of achieving the desired effect. If a patient goes into cardiac arrest and CPR is initiated, it may be quantitatively futile to continue if the patient remains in asystole after several minutes. The probability of achieving the goal, namely restoring breathing and circulation, is minimal.
3. **Qualitative** futility is when the intervention, even if successful, will produce such an undesirable outcome that it is best not to attempt it. In the above example, doctors may decide that, even if the patient's breathing and circulation are restored after 30 min of CPR, the extent of the neurological damage will be such that the patient's quality of life will be unacceptable.

This distinction highlights the complexity of the term.[6] When is the likelihood of success so low that an intervention is quantitatively futile? One chance in ten? One in fifty? One in a hundred? When is an outcome so undesirable that an intervention is qualitatively futile? And who should decide?

In light of the potential confusion, it may be clearer to talk of the harms of an intervention outweighing the anticipated benefits (Jonsen et al. 2010). This would also avoid the negative connotations associated with the word 'futility', which suggests giving up and abandonment (Sokol 2009; Jonsen et al. 2010).[7]

1.1.1.2 Best Interests

The same point about the illusory nature of certain terms applies to 'best interests', which can be used in a deceptively simple manner: "we are going to operate because it is in the patient's best interests". Like futility, the term contains

---

[6] Futility arguments can be divided into two parts: a *factual prediction* (for example, that the chances of success are 1/50, or that the patient will be unable to interact with others), and an *evaluative judgement* (that a 1/50 chance of success, or the inability to interact with others, is sufficient justification to withhold treatment).

[7] A complete version of my *BMJ* column on futility is reproduced in Appendix 4.

subjective, evaluative components. One person's best interests, such as a life prolonged but plunged in silence and darkness, can be another person's hell.

The Mental Capacity Act 2005 in England and Wales, which is concerned with persons lacking capacity, contains no definition of 'best interests'. Instead, it states that 'the person making the determination must consider all the relevant circumstances' (MCA 2005, s4(2)).

More helpful is the checklist of considerations which appears in the Act. The decision-maker must, so far as is reasonably ascertainable, consider:

1. The patient's past and present wishes, and feelings (in particular, any written statement);
2. The beliefs and values which are likely to have influenced the patient's decision if he had capacity; and
3. Other factors which the patient would probably have considered if able to do so. (s4(6))

It is clear that the Act attempts to respect the patient's autonomy, even though the patient is currently insufficiently autonomous to make medical decisions. Insufficient autonomy, however, does not mean no autonomy. The Act states that the decision-maker 'must, so far as reasonably practicable, permit and encourage the person to participate, or to improve his ability to participate, as fully as possible' in the decision (s4(4)). In practice, this means using clear, simple language, conveying information in manageable chunks, perhaps using visual aids such as photographs and illustrations, and talking to the patient at an appropriate time and location.

In determining best interests, the Act requires the decision-maker to consult the views of others, including 'anyone named by the person as someone to be consulted on the matter in question or on matters of that kind', and 'anyone engaged in caring for the person or interested in his welfare' (s4(7)). Of course, in practice, this may not be possible. The situation may be urgent and the patient unbefriended. There may not be time to seek the views of others. The Act thus qualifies the above with 'if practical and appropriate'.

Once the relevant circumstances, and the views of the patient and others, have been considered, the decision-maker must weigh up all the factors and determine what, in the circumstances, is in the 'best interests' of the patient. Determining a patient's best interests is no exact science.

1.1.2 Asking the Right Questions

Darwin Ortiz, an esteemed magician and theorist of magic, tells magicians that if they can get spectators to ask the wrong question about how the trick was done, they can guarantee that the spectators will never find the solution (Ortiz 2007).

Imagine a magician who signs a coin with a permanent marker and vanishes it from his hand to make it reappear in his pocket. A spectator trying to find a solution will consider ways to secretly transfer the coin from hand to pocket.

Perhaps he held it between his fingers? Or on the back of his hand? A fellow magician, however, will note that the coin was signed by the magician, not the spectator, raising the possibility that the magician used two different coins bearing his signature. The coin that 'appeared' in his pocket may have been there all along, pre-signed. There was no transfer. Without asking why the magician signed the coin himself, the spectator will search in vain for the solution.

Some clinicians presented with an ethical problem digress into more comfortable but quite irrelevant territory, losing track of the ethical issue. An anaesthetist may question the type of bolus used in an operation even though it bears no significance to the ethical problem.[8] A good ethicist, on the other hand, will cut straight through the irrelevant features of the case to the nub of the problem. Asking the right questions opens up the ethical dimensions of the case.

One way to focus the mind on the ethical issues is to use an ethics 'checklist'.

1.1.3 The Ethics Checklist

The idea for the checklist arose from time spent on ward rounds in the intensive care unit of a large hospital in Washington D.C. The checklist, in the form of a stamp imprinted on the patient notes, appears below (Fig. 1.1). The clinician simply ticks the boxes that apply to the case at hand. Take the real case of a competent, post-operative patient who, in spite of an inoperable leaking artery and a life expectancy of a few days, believes he is getting better. His relatives disagree over whether to tell him the truth about his prognosis. The completed checklist might look like this:

In the ward round, the presenter might announce "In terms of ethics, the patient appears unaware of his very short life-expectancy, raising an urgent disclosure issue ['disclosure']. The issue is further complicated by the disagreement within the family about the appropriateness of disclosure ['disagreeing relatives']. No end-of-life plan has been discussed with the patient, whose wishes are unknown, and the patient has no advance directive ['end of life issues' and 'patient wishes unclear']. The patient is currently for resuscitation ['end of life issues']." This would trigger a discussion about the ethical—and possibly legal—concerns which in some circumstances will lead to a formal ethics assessment, perhaps in a departmental meeting or the hospital's clinical ethics committee. In the case

---

[8] Nancy Sherman, a military ethicist, recounts a visit to Guantánamo, in which she and several senior military and civilian doctors were told that the seven detainees on hunger strike did not resist force feeding (Sherman 2010, p. 145). The commanding doctor, a Captain, showed the nasogastric tube used for feeding and described the use of anaesthesia and lubrication prior to insertion. Sherman remarks: 'Not one physician asked about the consequences of not acquiescing to the insertion of the tube; none openly worried that acquiescence might not be the same thing as consent; none voiced the concern that pulling out a nose tube funnelled down the back of one's throat to the top of one's stomach might, in some circumstances, be painful'. She concludes that 'medical and technical talk about equipment displaced responsible moral discourse about care.'

Fig. 1.1 Ethics checklist

Ethics Checklist	
Patient's wishes unclear/refusal of Rx	✓
Questionable capacity to consent	
Disagreeing relatives	✓
End of life issues (DNR, adv. dir., LPA)	✓
Goal of care/ appropriate Rx?	
Confidentiality or disclosure issue	✓
Resource or fairness issue	
Other (please note)	
No notable ethical issues	

above, a meeting with the relatives and clinicians should be called urgently to address the matters of truth-telling and advance care planning.

The checklist serves the dual function of identifying ethical issues and prompting a discussion on how best to resolve them. It does not provide a solution, but constitutes the first step in a process. The process may end at once if there are no notable issues, at least until the next ward round. Even in intensive care, not all cases will raise *notable* ethical issues (there is such a thing as ethical hypochondria, characterised by a belief that every encounter with a patient raises a profound ethical problem). Alternatively, the process may continue for days, weeks or even months for a complex case.

A value of moral perception is the prevention of potentially explosive ethical problems through early recognition and resolution. Some call this 'preventive ethics', in contrast to 'reactive ethics' (McCullough 2005). Ethically astute clinicians can sniff the faintest odour of an ethical problem and take preventive measures to minimise the chances of an eventual stench. This may be as simple as getting relatives involved in decision-making earlier than would otherwise be the case. It is well known that early discussions can avert a crisis (Fins 2006, p. 84). For those clinicians not yet blessed with a sensitive ethical nose, the checklist can assist in early identification of ethical problems and the development of greater ethical perception.

The checklist is not copyrighted, and readers who see in it some practical value are free to adapt it to their own specialty, and to pilot it. It is as good a research project as any. One version of the checklist is already in use in Washington Hospital Center in Washington D.C.

1.1.4 Post Mortem

Serious chess players conduct 'post mortems' after their games. They analyse their moves and identify which were good and which could be improved. In medicine, the morbidity and mortality meeting serves a similar function, but it is rare to hear any mention of possible improvements on the ethical front. Asking questions such as "how did we handle this case?" and "can we do it better next time?" may reveal ethical issues that were poorly handled, or that went undetected. Next time, hopefully, these hitherto invisible issues will be apparent to all. As the philosopher Lawrence Hinman has written, 'learning to see is an essential part of learning to be wise' (Hinman 2000, p. 413).

These questions about ethical practice, and their answers, can form the basis of an article or a presentation (see Chap. 2). Case reports need not be confined to purely medical matters. An ethics case report can contain valuable lessons for clinicians in other institutions and can help raise the ethical standards in the specialty.

1.2 Step 2: Moral Reasoning

Now, perhaps, you can see better. You are more alert to ethical issues, even in 'medical' mode; you are more cautious of words such as 'best interests' and 'futility' and the assumptions hidden within; you ask pertinent questions about the ethics of a case; you may have internalised a checklist of common ethical issues in your practice; and you regularly reflect on past dilemmas, accruing insight and experience.

Yet, identifying an ethical problem is only the first step. Deliberation must follow, and this is where an ethical framework is most helpful. Former world chess champion Gary Kasparov instructed chess players to 'become intimately aware of the methods you use to reach your decisions' (Kasparov 2007, p. 11). The same advice applies to ethical decision-making. Using a framework promotes structured, complete, and transparent reasoning.

There are dozens of ethical frameworks, and each ethicist will have his preferred one. I present two, both of which I have used in the hospital context under time constraints, in clinical ethics committee meetings, in the writing of formal reports, and in academic analysis. Which you use, if any, is a matter of personal preference.

1.2.1 The Four Principles Approach

The first framework will be familiar to many readers. It requires the application of four moral principles to the problem at hand. The four principles are **respect for autonomy, beneficence, non-maleficence** and **justice** (Beauchamp and Childress 2009). A brief description of each principle follows.

1. Respect for autonomy

Literally meaning 'self rule', autonomy refers to people's ability to make choices for themselves, based on their own values and beliefs. In medicine, the principle requires clinicians to, amongst other things, respect a competent patient's deliberated wishes and to provide sufficient information to help patients make informed decisions. Note that the principle is that of 'respect for autonomy', not 'autonomy'.

2. Beneficence

This principle requires clinicians to act in the best interests of patients. Since what counts as a benefit (and a harm) may differ from person to person, this principle is linked to the principle of respect for autonomy. If you respect someone's autonomy, you are more likely to benefit them, as judged by that person's own view of what constitutes a benefit. In medicine, this principle goes hand in hand with the principle of non-maleficence and should be considered in conjunction with it (see below).

3. Non-maleficence

This refers to the clinicians' moral obligation not to cause harm to patients. However, nearly all attempts to help patients—through drugs, procedures, and even words—carry a risk of harm. For this reason, it is best to describe the principle of non-maleficence as the obligation to avoid causing *net* harm to patients. *Primum non nocere* ('above all, do no harm') is inaccurate. A clinician can legitimately inflict harm in some situations, as when drilling a hole in a patient's skull, as long as the harm is outweighed by the benefits. It would be more precise, though less pretty, to say *primum non plus nocere quam succurrere* ('above all, do not harm more than succor') (Sokol 2008).[9]

4. Justice

The principle of justice is probably the most complex of the principles. It refers to a collection of obligations which includes the obligation to act fairly, to distribute resources justly, and to respect people's human rights (rights-based justice) and the laws of the jurisdiction (legal justice).

The principles are broad in scope, but more specific rules can be derived from each principle. The process of going from the general principles to more useful, action-guiding rules is called *specification*. Hence, under respect for autonomy, are rules such as 'obtain consent', 'respect confidentiality' and 'tell the truth'. Under beneficence and non-maleficence are rules such as 'acquire relevant skills' and 'keep up to date' (i.e., maintain your skills), and under justice are 'respect the law', 'respect human rights', and 'abide by your professional code'.

---

[9] The article appears in full in Appendix 5.

A 'principlist' will scan the issues in a case and sort them under each of the principles. Thus, if a patient in urgent need of treatment arrives by ambulance following a road traffic accident and is unable to give informed consent, the principle of beneficence provides a *prima facie* obligation to treat the patient. The principlist may need to reflect on the principle of justice if there are many such patients or, alternatively, if the treatment is so costly or scarce that it is likely to deprive others of medical care. Non-maleficence may play a part in the ethical calculus if, for example, the treatment is likely to be more harmful than non-treatment. Respect for autonomy would come into play if there was some indication that the patient had made an advance decision not to receive treatment in such circumstances.

If each of the four principles represents a single voice, there will be times where a sweet melody will be heard at the conclusion of your analysis. Each voice of the quartet will sing in harmony. At other times, there will be dissonance. The principles may—and often do—conflict with one another, and an obligation derived from one principle, such as 'respect patient confidentiality', may be trumped by more compelling obligations, such as 'respect the law which requires notification of certain infectious diseases' and 'prevent serious harm to others'. Each principle is not absolute, but *prima facie*. In other words, it is binding unless trumped by a stronger principle.

Whether a principle is stronger than the other will depend on the facts of the situation. There is no fixed hierarchy of principles. The same is true in medicine. The ABC doctrine of airway, breathing and circulation generally holds true, but it is not an absolute rule. In the combat environment, the prevention of massive haemorrhage takes priority over airway management, changing the mnemonic from ABC to CABC (the initial 'C' stands for catastrophic haemorrhage) (Hodgetts et al. 2006). The particular facts of the case provide the basis for establishing the relevance and weight of the moral principles. A firm grasp of the facts and context is therefore essential to a sound ethical analysis. "Always check the patient's notes" said Dr John Lynch, a clinical ethicist and oncologist, on my first day as a Visiting Scholar at Washington Hospital Center. If you want to know the facts of the patient's case thoroughly, you must consult the notes. Become the patient's medical biographer.

In my early days as a hospital ethicist, when faced with complicated ethical problems, the four principles regularly dispersed the mists of doubt and ignorance to reveal blue sky. Even with numerous and overlapping issues, the four principles provided a comforting starting point, and when the time came to present my thoughts more formally, they structured my analysis.

Consider another example. In a pandemic situation, a patient's relative asks that one of the few ventilators be given to their gravely ill and unconscious relative. The main tension, using the principles, is between, on the one hand:

• respecting the autonomous request of the relative, and
• benefiting the patient medically by providing the ventilator
and, on the other hand:

- duties of beneficence and non-maleficence towards other patients, both current and future, who may derive greater benefit from the ventilator than the patient, and
- an obligation of justice to use limited resources effectively.

The principles do not provide an answer to the problem, but they greatly facilitate the 'opening up' of a case into its constituent parts.

The second—and usually more difficult—step in the four principles approach requires carefully "weighing up" (or evaluating) the various principles and values identified in the first step, and arriving at a well-reasoned choice. However, you balance the conflicting principles, you must be in a position to justify why one principle has priority over another. This is called 'deliberative balancing' (DeMarco and Ford 2006). Otherwise, your decision will appear arbitrary.[10]

Consider a case in which the parents of a young girl are refusing life-saving treatment for her on religious grounds (DeMarco and Ford 2006). You have identified some key values such as parental autonomy, the avoidance of harm, the burdens of treatment, and so on. In a simple case, the treatment is not burdensome and the benefits considerable. In your analysis, you would argue that the principles of beneficence and non-maleficence trump respecting parental autonomy because the harm to the child is very high and the burdens of the treatment are minor. The benefit to the child is sufficiently large to justify overriding the parental rights of the parents. If, however, the burdens of treatment were higher, the potential benefit much lower, and the young girl refused the treatment, then the conclusion may well be different.

A lawyer in court must be prepared to answer a question from the judge asking "How have you come to this conclusion?". The lawyer must also anticipate how the opponent will try to undermine his argument, and how he would respond to the attacks. When preparing the case, he must consider how he would argue if on the other side. Similarly, a good ethicist will look for any holes in his own reasoning and anticipate opposing views. Be harsh on yourself. Ask for the opinion of trusted friends and colleagues and instruct them not to hold back.[11] It will improve your reasoning by providing different perspectives and may detect errors or unsupported assumptions. Important ethical decisions made without consulting others should be rare indeed.

Of course, the search for supporting reasons will, if pushed, only take you so far. If some irksome soul were to keep asking you "why?" at each answer you give, you will eventually have to concede that there is no reason other than intuition. The philosopher Ludwig Wittgenstein gives the analogy of digging with a spade:

---

[10] Even seasoned judges can, on occasion, make arbitrary decisions. In 1981, an Old Bailey judge spared a defendant prison for a criminal offence: "you have caught me on a good day", the judge announced, "I became a grandfather this morning again." (Pannick 1987, p. 18)

[11] You must, however, maintain patient confidentiality.

If I have exhausted the justifications I have reached bedrock, and my spade is turned. Then I am inclined to say: "This is simply what I do." (Wittgenstein 1963).

If you balance conflicting rules or principles based on intuition rather than reasons, this is referred to as 'intuitive balancing'. Clearly, it is preferable to support your decisions with well-deliberated reasons, but that is not always possible. DeMarco and Ford argue that when there is no clear winner as to which value or obligation overrides the other, intuitive balancing may be appropriate (DeMarco and Ford 2006).

Finally, a word of warning. You will hear some ethicists criticise the four principles. They will say that they are vacuous, simplistic, reductionist, or boring. The framework is taught in many medical schools and it can hardly be denied that it is often applied poorly. The weaker medical students simply justify their conclusion by referring to one of the principles ("the doctor should continue aggressive treatment because of the principle of beneficence") and that will be the extent of their moral reasoning. They fail to identify the competing principles, conveniently ignore relevant facts and the 'bigger picture', and say nothing about why their chosen principle should prevail over the others. However, the fact that a framework can be misused does not mean it is useless. The four principles can be used with great virtuosity. With practice, and in the right hands, it is a most valuable instrument.

1.2.2 Applying the Four Principles

The four principles can be used to survey the ethical landscape of a particular specialty. Table 1.1 below is a typology of ethical issues in surgery (Adedeji et al. 2009).

Readers will note that some of the ethical issues are covered by several principles. The duty to possess adequate technical skill ('surgical competence') falls under beneficence and non-maleficence. In fact, it could also be included under justice, as the law requires practising surgeons to exercise reasonable skill and care.[12]

---

[12] Some readers will have heard of the *Bolam* test for negligence. Lord Scarman, in his judgment in *Sidaway v Governors of the Bethlem Royal Hospital* [1985] AC 871, provides a succinct formulation of the *Bolam* test: 'A doctor is not negligent if he acts in accordance with a practice accepted as proper by a responsible body of medical opinion'. However, the courts will not accept unquestioningly the opinion of a professional body of doctors on what counts as proper practice. It must also have a logical basis (*Bolitho v City & Hackney Health Authority* [1998] AC 232).

For lawyers, proving that a doctor, or a nurse, was negligent (i.e., breached a duty of care) is not enough to succeed. They must also show that the negligent act or omission *caused* the patient's injury or loss, and that the type of harm was a foreseeable consequence of the act or omission. For an introduction to the law of tort, of which clinical negligence forms a part, see Bermingham and Brennan (2008).

Table 1.1 Typology of ethical issues in surgery

Ethical principle	Ethical issues in surgery
Respect for autonomy	Informed consent for surgery
	Truth-telling (to patients, relatives, and colleagues)
	Consent for involvement of trainees in surgical procedures
	Confidentiality
	Respecting patient's requests (for procedures/particular surgeons)
	Good communication skills
Beneficence	Surgical competence
	Ability to exercise sound judgement
	Continuous professional development
	Research and innovation in surgery
	Responsible conduct
	Functioning equipment and optimal operating conditions
	Minimising harm (including pain control)
	Good communication skills
Non-maleficence	Surgical competence
	Continuous professional development
	Ability to exercise sound judgement
	Recognising the limit of one's professional competence
	Research and auditing
	Disclosure and discussion of surgical complications, including medical errors
	Good communication skills
Justice	Allocation of scarce resources
	Legal issues
	Respecting human rights
	Whistleblowing

'Good communication skills' also spans principles. It is well known that poor communication, such as insensitively breaking bad news, can cause distress (non-maleficence) and loss of trust, and lead to avoidable tensions and disagreements between the patient, family and medical team. At times, failing to engage meaningfully with the patient, to elicit his wishes or fears, can result in inappropriate treatment decisions. Good communication is more than 'being nice' or polite. It requires effective and ongoing communication between the patient, the family and the medical team.

A similar typology can be drawn for any medical specialty, and if such an overview has not been conducted in your specialty my advice is to grab the opportunity and publish it in a journal.

As shown earlier, the four principles can also be used in the context of an individual case. The main steps are as follows:

1. **Identify the key facts of the case**. Good ethics starts with good facts. In some cases, important facts will be missing. Do your best to obtain them. Read the patient notes carefully. Talk to colleagues and the patient's relatives. If the facts

are unavailable, then proceed in the knowledge that relevant facts are lacking and be prepared to modify your analysis in light of new information.

Some medical students cannot tolerate the uncertainties and ambiguities inherent in many ethical dilemmas. They cover their ears at the dissonance of the principles' voices. They get frustrated at the lack of a definite prognosis, or the impossibility of obtaining the patient's views. Rare is the case where all the requisite facts are available. It is important, for this reason, to develop a tolerance for uncertainty. There is little point in bemoaning the inevitable.

2. **Apply the principles in turn**, identifying the moral obligations derived from each principle. If paper is at hand, write 'respect for autonomy', 'beneficence and non-maleficence' and 'justice' as broad headings and jot down notes under each heading. Do not limit yourself to the patient. Respecting the autonomy of the patient is important, but the autonomous wishes of the relatives and even the healthcare professional may be relevant in some cases. When applying the principle of beneficence, consider the benefits to the patient *and* the potential benefits to others, such as relatives, other patients, future patients, and the medical team.

The same applies for harms. Respecting the confidentiality of a patient may benefit the patient by respecting their autonomous wish not to have their secret divulged, but respecting the privacy rights of that patient may also benefit *future* patients by maintaining the all-important commitment to patient confidentiality. There is, after all, a strong public interest in clinicians maintaining confidentiality. If this commitment to confidentiality is weakened, future patients may be reluctant to share private but medically-relevant information, with potentially adverse effects on their medical care.

A case may be easy because the principles do not conflict with each other, or because one principle is indubitably dominant in the circumstances. However, if the case is complicated, your analysis should reflect the complexity. Though tempting, do not pretend that a case is simple when it is not. The analysis should expose the complexity in a clear and structured way.

3. **Highlight the tensions between ethical principles, and seek ways to resolve them**. Think outside the box. Wise solutions to problems are often creative ones. You may have heard the story of the professor who wanted to research the problem-solving skills of chimpanzees. He put a chimp in an empty room, and suspended a banana from the ceiling, just out of the chimp's reach. He then placed several crates around the room. Would the chimp collect the crates, and stack them to reach the elusive banana? As the professor was positioning the crates, the chimp waited patiently in the room. When the professor walked directly below the banana to reach the other side of the room, the chimp jumped on him, stood on his shoulders, and grabbed the fruit. An easy and unexpected solution (Gardner 1978, p. 6)!

On several occasions, I have seen seemingly intractable disagreements between clinicians and families resolved when one bright spark suggested that the hospital chaplain talk to the family in private. The distrust of the families eased, and meaningful discussions with the medical team followed.

4. If no easy solution is found, **evaluate the pros and cons of each realistic course of action**. What are the consequences of each option, in particular the risks, burdens and benefits? How likely and significant are these and how do they relate to the patient's values and wishes? What professional, legal and moral duties are in play? Are any of the options in breach of the General Medical Council guidelines, for example? Resolving some ethical problems will require document-based research, such as consulting the guidance of the British Medical Association, although this may not be possible if the problem is very urgent.

There are two additional points to note. First, some people, to get their way and stifle deliberation, may try to persuade you that the matter is urgent when it is not. If urgency is pleaded as a reason to shorten the deliberative process ("we don't have time to contact the patient's GP") make sure the decision is truly urgent. Beware of decisions driven by convenience rather than sound judgment. As Lord Atkin wrote, 'convenience and justice are often not on speaking terms' [*General Medical Council v Spackman* (1943) AC 627 at 638].

Second, an ethical problem may not have a single 'right' answer. There may be compelling reasons to do act X and to do act Y, either of which could be justified using the four principles. Just as it may be quite acceptable to either clip or coil a cerebral aneurysm, several good solutions to an ethical problem can co-exist. Note that this is not the same as saying that there are no right

Table 2 The four quadrants in order

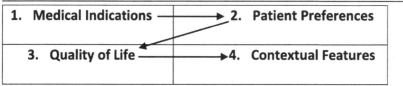

1. Medical Indications	2. Patient Preferences
3. Quality of Life	4. Contextual Features

answers in ethics (an assertion often made by medical students and even some clinicians). There are right answers in ethics, although sometimes more than one, and of course there are many bad answers.

5. **Review the decision**. Articulate the decision and make the reasoning explicit. Is it rigorous and defensible? Anticipate counter-arguments. If time allows, share your conclusion and its rationale with colleagues, and review the decision if necessary.[13]

1.2.3 The Four Quadrants Approach

This second approach to ethical case analysis is particularly popular in the United States, though less well known in Europe. It was developed by Jonsen et al. in their 1982 book *Clinical Ethics* (Jonsen et al. 2010). It consists of four quadrants or topics: medical indications, patient preferences, quality of life, and contextual features. Unlike the four principles, they should be addressed in a precise order (Table 1.2):

1. **Medical indications**

This first quadrant focuses on the clinical facts. It requires reviewing the medical situation, identifying the various treatment options, and examining how *medically* to benefit the patient with minimum harm. What is the patient's diagnosis and prognosis? Is the condition reversible? Whether a problem is chronic, acute or reversible can have ethical relevance. A patient with end-stage metastatic cancer and multi-organ failure who lapses into cardiac arrest is a quite different proposition to a one-off acute arrest on the operating table.

The goals of any proposed treatment and the likelihood of achieving these goals are important considerations. Is the goal to cure the patient of the disease, to maintain or restore a particular function, to improve quality of life, to prolong life for a certain period, to ensure a good death, or even to satisfy the patient's

---

[13] For a detailed application of the Four Principles approach to a genetics case, see Raanan Gillon's (2005) chapter 'Families and genetics testing' in Ashcroft et al. 2005, pp. 165-86.

relatives? Disagreements over the goals of treatment can generate conflict, including within the medical team.

I remember a case involving an extremely sick woman who had lost all decision-making capacity. The surgeon, who had performed the previous eight operations on this patient, wanted to try one last-ditch operation, while the physician and the nurses believed a palliative approach was best. The relationship between the surgeon and physician had degenerated to such an extent that the nurses called the hospital ethics team. They simply could not agree on what was the appropriate goal. This shows that determining what is medically best for a patient is not always uncontroversial, and that clarifying the goals of treatment can be a vital step towards resolving the ethical problem.

This first quadrant requires a preliminary conclusion on what is medically indicated for the patient. 'Preliminary' because it is subject to change depending on the evolving medical situation and consideration of the three remaining quadrants. No final decision should be made before all the quadrants have been examined.

2. **Patient preferences**

The second quadrant embodies the principle of respect for autonomy. It focuses on what the patient wants or, if unable to express a view, what he would have wanted in this situation.

The first step is to establish if the patient is capable of making an autonomous decision. Avoid jumping to conclusions about capacity. Physical disability, mental illness, religious fervour, personal appearance, alcohol or drug use, homelessness, or a plainly bad decision (such as refusing a treatment that is clearly beneficial and with few side-effects) do not entail incapacity. If in doubt, assess the patient or refer to a psychiatrist, but do not automatically assume incapacity. If the patient has capacity, there will be a strong, though not necessarily determinative, argument in favour of respecting the patient's autonomous wishes.[14]

The patient's values and preferences may either confirm or change the treatment goals identified in the first quadrant. If the medical team decided that an intervention was indicated to prolong life by several months (goal 1), but the competent patient expressed a reasoned preference for a shorter life without the burdens of treatment (goal 2), then goal 1 should be reconsidered in light of this. Clearly, the patient's view should not be accepted unquestioningly. The medical team should make sure that the patient is aware of the implications of his choice.

A useful acronym relating to the provision of information to patients is PARQ. It stands for Procedure (what it entails), Alternatives (including doing nothing), Risks (of the procedure and relevant alternatives), and Questions (invite the patient to ask questions). In some hospitals in the United States, the clinicians write

---

[14] Note that capacity is task-specific. A person may have capacity to decide in which arm to have an injection but not to decide whether to undergo an operation for their acoustic neuroma. "Does this patient have capacity?" is more accurately rephrased as "Does this patient have capacity to make this particular decision?".

'PARQ' in the patient notes to show that they have discussed these elements with the patient.

If the patient does not have capacity, evidence of past wishes and relevant values, whether in the form of an advance statement or the credible account of relatives or the patient's general practitioner, can provide helpful guidance on the patient's likely preferences.

3. Quality of life

The third quadrant examines how a proposed intervention will affect the patient's quality of life. If the treatment works, what will the patient be able to do? What physical, mental and social deficits will there be? Would the patient's anticipated quality of life be acceptable or would life be so grim as to be a curse? Would a less aggressive approach lead to a better quality of life? There is, inevitably, a subjective component to the evaluation of this quadrant, but it must be addressed nonetheless.[15] Few of us would value a life deprived of all quality.

John Lantos and William Meadow, in their short book *Neonatal Ethics* (Lantos and Meadow 2006), break down the concept of quality of life into four components:

1. Anticipated cognitive or cerebral function;
2. Anticipated physical disabilities;
3. Pain and suffering associated with the disease;
4. Burdens of future treatment.

This classification allows us to be more specific when talking about the 'quality of life'. We can now support a vague statement such as "this patient's quality of life will be unacceptable" with some form of reasoning: "As a result of trisomy 18, this baby will have virtually no cortical function and will be unable to walk, talk, or carry out simple activities", or "As a result of severe epidermolysis bullosa, this patient will be in acute pain, and may be poorly responsive to pain control", or "As a result of her advanced motor neurone disease, long-term mechanical ventilation will be extremely burdensome for this baby". Lantos and Meadow note the value of their classification, while acknowledging that it does not identify the point at which quality of life becomes unacceptable:

> By breaking down the concept of quality of life into subcomponents, it becomes possible to analyze which elements are driving the decision. [...] In each of these areas, there are no bright-line distinctions between acceptable and unacceptable quality of life (Lantos and Meadow 2006, p. 81).

---

[15] Harvard psychologist Daniel Gilbert remarks that people consistently overestimate how much, and for how long, negative events such as losing a job or breaking up a relationship will affect them (Gilbert 2006, p. 153). People, in short, are more resilient than they think. This observation also applies to illness and disability. Menzel et al. note that 'Chronically ill and disabled patients generally rate the value of their lives in a given health state more highly than do hypothetical patients [who are] imagining themselves to be in such states.' (Menzel et al. 2002). This calls for caution in projecting our evaluation of quality of life onto others.

4. Contextual features

The last quadrant encompasses legal, cultural, familial, religious, economic and other issues left untouched by the other quadrants. Questions that would usually be asked under this quadrant include "what does the law require?" and "what does the General Medical Council or British Medical Association guidance require?" So, if a patient lacks capacity and has not made a legally binding advance decision, the medical team should, under the Mental Capacity Act 2005, act in the best interests of the patient. Failing to address contextual issues may lead to a crude, incomplete, or misguided ethical analysis. If in doubt about the law, it may be wise to contact the institution's legal department or your defence organisation. In England, the GMC and the BMA have ethics hotlines.

Issues of resource allocation and fairness fall under this quadrant, as well as any biases or prejudices of stakeholders that may influence a decision. I have seen clinicians, emotionally bound to long-term patients and unable to "let go", over-treat patients. Challenging patients, such as those who repeatedly self-harm or create trouble on the ward, can also adversely affect the reasoning of their exasperated clinicians.[16] These unpopular patients, in contrast, can be under-treated. Team meetings, in which members of the multi-disciplinary team are involved in the decision, can reveal and off-set the often subtle prejudices of some individuals. Such meetings have the added benefit of keeping the whole team up to date about the care plan.

The impact of a medical decision on the patient's relatives and on other persons will also be captured by this quadrant. The views of a patient's family may not hold the same moral weight as the wishes of the patient, at least in the United Kingdom, but this does not mean that they should be ignored altogether. No man is an island, as the poet said.

Astute readers will have noticed that this quadrant is broader than the others, and examines the situation from a higher vantage point. It is, in truth, a hotchpotch of features, and not all of them will be relevant to any single case, but it would be foolish to skip it. At least *some* features will be pertinent to any case.

1.2.4 Applying the Four Quadrants Approach

Consider a case based on an example by Jonsen and others (Jonsen et al. 2010). Luke is a 7-year-old boy with acute myeloid leukaemia. After a course of chemotherapy, he

---

[16] In a recent article, psychiatrist Dr Stephen Peterson wrote: 'Physicians can become frustrated when patient are hospitalized repeatedly for conditions linked to or exacerbated by their obesity. This can cause them to maintain an emotional distance from their obese patients. The patient's physical appearance can produce feelings of distaste, if not outright revulsion' (Peterson 2011, p. 6).

relapses and obtains a bone marrow transplant from his sister. Unfortunately, he relapses soon after the transplant. The oncologist advises Luke's parents against further chemotherapy, but the parents insist on it. A course of experimental chemotherapy is attempted but unsuccessful. Luke, a formerly vivacious boy, is despondent. He asks "why do I have to keep going on with this?".

1.2.5 Medical Indications

We should avoid the temptation of plunging straight into the ethical dilemma. The first step is to clarify the medical situation. Clinical uncertainty can lead to moral uncertainty. What is the likely prognosis with and without further aggressive treatment? What is the goal of treatment, and the likelihood of achieving it? What if the treatment doesn't work? What are the benefits and the harms of any proposed intervention?

Sometimes, what should be done from a medical perspective is straightforward. At other times, the medical situation is more complex, and clinicians will disagree over what is medically indicated. Recall that the conclusion of this quadrant is always a provisional one, subject to re-evaluation in the light of the other quadrants yet to be explored.

1.2.6 Patient Preferences

We must find out if Luke can make decisions about his care. Does he understand the situation? If so, what are his views about it? Although Luke's views do not, from a legal perspective, hold the same weight as an adult, they should nonetheless inform the final decision. Furthermore, leaving him out of the picture may result in him feeling abandoned or isolated. Note that if Luke was an adult lacking capacity, the quadrant would require seeking any prior preferences.

1.2.7 Quality of Life

If further aggressive treatment is instigated, what impact is it likely to have on Luke's quality of life? How will Luke's mental, physical and social well-being be affected? These questions are important to ascertain what is in Luke's best interests. If further aggressive treatment is medically indicated, this quadrant asks "what will be the quality or value of the additional weeks or months?"

Fig. 1.2 DNR tattoo

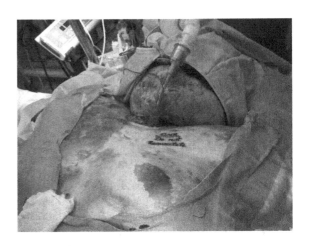

1.2.8 Contextual Features

Here, we would explore the views of Luke's parents and possibly his sister. We would examine any relevant religious or cultural issues. We would seek guidance published by relevant bodies, and consider any pertinent resource allocation issues.

Enthusiastic readers may wish to analyse the case above using the four principles. They will discover that there is no incompatibility between the two frameworks. The principles of beneficence and non-maleficence are linked to 'medical indications' and 'quality of life', the principle of respect for autonomy with 'patient preferences', and justice falls within 'contextual features'.

1.2.9 A More Detailed Example

Below is a fuller analysis of a real, and quite extraordinary, case (Sokol et al. 2011).

The patient was a 22-year-old woman, who was admitted to a District General Hospital with an overdose of painkillers and antidepressants. She had a Body Mass Index of 51 and a history of self-harm. On several occasions, she had swallowed foreign bodies which required surgical removal. Minutes after discharge from the emergency department, still in hospital grounds, she doused herself in lighter fluid and set herself alight. She was found by nearby paramedics and readmitted to the emergency department with burns to the head and neck. She was intubated and admitted to intensive care, and then transferred to a specialist burns unit.

In the burns unit, the burnt skin was removed and it was during this process that a message was found. Tattooed on the patient's chest, in a prominent position, was the following: DNR (underlined) Do Not Resuscitate (Fig. 1.2). The patient required further resuscitative measures to survive. Should the medical team initiate those measures?

The order of analysis is:

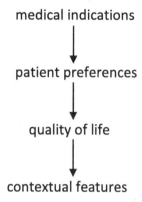

medical indications

↓

patient preferences

↓

quality of life

↓

contextual features

1.2.10 Medical Indications

The patient has a difficult airway caused by swelling, but the medical indications are not in doubt. She needs ventilation, fluid resuscitation, and enteral feeding.

The immediate medical problems are acute and reversible. The goals of treatment are: cure, restoration of function, and prolongation of life. The likelihood of achieving these goals is high.

If thinking in terms of the four principles, the principles of beneficence and non-maleficence suggest that, from a *medical* perspective, the benefits of instigating urgent resuscitative measures outweigh the harms.

1.2.11 Patient Preferences

In this case, the patient is not mentally capable of making an autonomous decision. There are no indications of past wishes: no formal advance decision, no relatives at hand, and no information from her general practitioner.

We cannot derive a clear idea of her preferences from the tattoo alone. We do not know the circumstances in which she had the tattoo. Was she lucid or delirious at the time? Drunk or drugged? Was it a dare or a joke? Was it etched last month or six years ago? The very fact that she set herself alight in the hospital grounds suggests she may not have intended to kill herself. It may have been a cry for help.

The truth is that we do not know her autonomous wishes on resuscitation in this situation. The degree of certainty is not sufficiently high to justify withholding resuscitation and allowing her to die.

For the same reasons, the principle of respect for autonomy does not point strongly in favour of non-resuscitation.

1.2.12 Quality of Life

Although we can infer that, at the time of attempting suicide (if that was indeed her goal), the patient was deeply unhappy, we do not know how she will feel if given another chance at life. Will she be grateful or resentful? Will her unhappiness continue or will this episode signal the start of a brighter outlook? Again, we cannot say for sure and, for this reason, it seems wise to err on the side of caution. We must assume that her quality of life will be acceptable.

This reasoning would also hold in applying the principles of beneficence and non-maleficence. We do not know if resuscitation is in her best interests, but we must assume that it is. With the information at our disposal, we must adopt the default position that people with reversible, life-threatening injuries would rather live than die.

1.2.13 Contextual Features

The main factor here is the law. As there is no legally binding advance decision, the medical team should resuscitate the patient if deemed in her best interests. The tattoo does not satisfy the stringent requirements of a valid advance decision, as stipulated by the Mental Capacity Act 2005. The tattoo is neither signed nor witnessed, and is not accompanied by a statement from the patient saying that it applies even if her life is at risk.

There is no legal obligation to comply with the directions of the tattoo. In fact, the principle of justice points more forcefully towards resuscitation than not resuscitating. It would be unfair to let her die, given the limited facts available. It would constitute abandonment, and possibly a violation of her right to life. This is a case in which the missing facts are as important as those we have.

The patient is a regular visitor to the Emergency Department and, by reason of her tendency to swallow objects, the operating table. She is what some clinicians call a "difficult" patient. We must be careful not to let these feelings, which sometimes lurk in our subconscious, colour our judgement.

1.2.14 Conclusion on the Case

If resuscitated, the patient's clinical outlook is good. Her quality of life, presumably low before the suicide attempt, may improve. The law is clear: the tattoo's instructions are not legally binding, and the medical team should act in her best interests. These reasons collectively outweigh the risk that the patient may not have wanted to be resuscitated.[17]

As mentioned earlier, deciding which framework to use is a matter of personal preference. I have found the four quadrants approach most useful in hospital-based clinical scenarios. I have also recommended the approach in ethics guidelines for military clinicians working in conflict zones. The clinical focus, the clear order of progression from quadrant to quadrant, and the self-explanatory topics are appealing features for many clinicians. The approach does not differ widely from clinical case presentations, with their familiar signposts of chief complaint, history of present illness, past medical history, and so on. With practice, the framework can be applied quickly, even under pressure.

The four principles approach has a broader scope of applicability, and can be used to analyse macro-level problems, such as policy issues. I therefore choose my analytic framework based on the nature of the problem: for individual cases in a clinical context, I use the four quadrants; for everything else, the four principles.

Both frameworks help identify and analyse ethical problems in medicine. They improve moral perception and provide a systematic means of reasoning about even the most complex of cases. "System, or as I shall term it, the virtue of method", wrote William Osler, "is the harness without which only the horses of genius travel" (Bean and Bean 1950). As well as helping to defend a moral position, the frameworks allow you to challenge the reasoning of others, which is useful when trying to persuade a colleague that his view is flawed.

The frameworks do not, however, eliminate the need for judgement or justification. When principles or duties conflict, the frameworks do not assert which should trump the other, or give indications on their respective moral weights. They are an aid to moral decision-making, not a moral panacea.

Often, the solution will emerge naturally as you apply the framework. But, when the resolution remains unclear, you will need to exercise judgement to determine which course of action is best in the circumstances. Remember that more than one morally acceptable solution may exist.

Finally, do not let an analytic framework get in the way of common sense. If your conclusion seems at odds with common sense, reconsider it carefully. You have probably gone wrong somewhere.

---

[17] This case may appear simple when ensconced in an armchair but, in the moments following the discovery of the tattoo, under time pressure and with the patient's apparent wishes plainly inscribed on the patient's chest, the medical team had doubts about what action was appropriate. After a discussion in the operating theatre, they decided to continue resuscitative measures. After the incident, the plastic surgeon contacted me to discuss the decision and to learn a method with which to analyse this and future cases. I recommended the four quadrants approach.

1.3 Step 3: Moral Action

You have, by this stage, identified the moral problem and reasoned your way through it. You have evaluated the various options, and decided on a course of action. The next step is to draw up an implementation strategy.

Imagine that you have decided, after careful moral deliberation, to inform a child of his terminal illness. That is not the end of it. Further questions must be asked: when is the best time to break the bad news? How should the disclosure be phrased? Who is the most appropriate person to tell him? Once the child is told, how should we support him psychologically? These duties, relating to how to implement a moral obligation, are called 'duties of manner' or 'adverbial duties', and they can be of considerable moral importance.

The philosopher Robert Audi writes:

> May we not judge a person to have acted wrongly because of something done (say) crudely, insensitively, or condescendingly? It might be a type of act that is permissible or even obligatory, say helping a patient to get into a high bed. There are ways to do this that are wrong, such as doing it resentfully, complainingly, or violently (Audi 2006, p. 178)[18]

The point, succinctly, is that if you are going to do it, do it right.

It is common for people not to do 'it' at all, despite knowing that it ought to be done. These people, for all sorts of reasons, fail completely to act. I earlier referred to some of the medical students in the BMJ study who did not identify any problem with conducting vaginal examinations on anaesthetised women who had not given consent (Caldicott et al. 2003). Other students had more sophisticated levels of moral perception and reasoning, but they did not act on the conclusion of their reasoning. One 4th year student, for example, said:

> You couldn't refuse [the consultant's offer to perform an intimate examination] comfortably. It would be very awkward, and you'd be made to feel inadequate and stupid (Caldicott et al. 2003), p. 99.

A good decision can be an unpopular one, and difficulties with moral action can arise when you suspect that doing the right thing will diminish your popularity or cause upset.[19]

I remember attending one committee meeting where the other six members, all middle-aged men, were close friends with each other. They embraced when they met. Soon after the meeting began, it emerged that the most eminent member of the group had a serious conflict of interest. He had much to gain from one outcome, and much to lose from another. It was quite clear that the committee would not be taken seriously as long as he was on it. To my surprise, 20 min into the

---

[18] I have seen consent, even for major surgery, obtained in such a hasty and perfunctory manner that the patient must have understood little of the intervention. Obtaining consent is morally obligatory, but it is possible to obtain it in violation of the adverbial duties to do so in a meaningful, patient-centred way.

[19] You may then face a 'moment of truth'. See Appendix 6.

meeting, no one had raised the issue. The elephant was roaming freely in the room. So I pointed to it, as diplomatically as I could, and to his credit the member agreed to remove himself from the committee. Heaven knows what he and the other members thought of me. Perhaps they were grateful, but more likely they thought me an ass. The most difficult part in this moral problem was not perception or reasoning but speaking up.

It is in such delicate situations that the rather old-fashioned notion of virtue enters the fold. Virtues, such as courage, can help moral agents act as they should. In fact, there is a class of ethicists, known as 'virtue theorists', whose analytic framework is based on virtues (Hursthouse 1999). In essence, they examine a problem by asking what a virtuous clinician would do in the particular circumstances. For reasons of space, I do not examine any of these frameworks here but suffice to say that virtues are compatible with the frameworks presented in this book. Virtues such as benevolence, discernment, compassion, integrity, and courage are important, amongst other things, to facilitate clear, unbiased reasoning in the analysis stage and to motivate moral action.

1.3.1 Clinical Ethics Consultation in Action

Clinical ethics consultation, in which one or more ethics 'consultants' provide advice or guidance to clinicians, is the epitome of applied clinical ethics. These consultants, who form part of the broader medical team, do ethics on a daily basis, sometimes at the bedside. Observing how they conduct their activities, from moral perception to action, is informative.

Washington Hospital Center, a 926-bed hospital in Washington D.C., has a team of four professional clinical ethicists, providing around 500 formal and informal ethics consultations a year. When contacted by clinicians about an ethical problem, often by telephone, the ethicists use an initial 'intake' form to note down the key facts and prompt important questions. The form, completed with details of a hypothetical patient, appears below (Fig. 1.3):

After the initial contact, the ethicist consults the patient's notes to obtain a clearer picture of the situation, including the medical history. A close reading of the patient's notes is an integral part of the fact-gathering process.

A formal ethics consultation is usually arranged between clinician and ethicist, and the conclusion of the consultation is recorded on another form, which is then placed in the patient's notes. An example of the 'ethics consult form', again with details of the same imaginary patient, appears below (Fig. 1.4):

The 'comments' are succinct and provide clear instructions to the medical team. While the ethicist may have used a framework such as the four principles or the four quadrants, the details of the analysis are not included in the form. The form presents the conclusion of moral deliberation rather than the reasoning itself. Its main purpose is to provide practical, ethically sound advice. It may also offer some legal protection for the hospital and the clinicians.

1.3.2 Practice Makes Perfect

There is no secret to honing your skills in ethical analysis. The more cases you encounter, the more time you spend thinking about them in a meaningful way, the more proficient you will become as an ethicist. So, if keen to improve, look out for ethical cases in your clinical work, join or create a clinical ethics committee, read articles and books on ethics,[20] and attend ethics conferences.[21]

---

[20] There is a growing number of medical ethics casebooks. Three books that I have used are Ackerman and Strong (1989) 'A casebook of medical ethics', Kuczewski and Pinkus (1999) 'An ethics casebook for hospitals' and, for surgical cases, Jones et al. (2008) 'The ethics of surgical practice: cases, dilemmas, and resolutions'. All three books are written by American authors, but many of the ethical issues also arise in UK hospitals.

[21] There is plenty of research showing that the highest achievers in sport, music, chess and many other fields are those who have spent the most time in solitary study (Syed 2010; Shenk 2010). What matters is not just practice, but *deliberate* practice. As Shenk explains, deliberate practice goes beyond hard work. It is 'practice that doesn't take no for an answer; practice that perseveres; the type of practice where the individual keeps raising the bar of what he or she considers success' (Shenk 2010, p. 56).

When I first read about the four quadrants approach, I took every opportunity to apply it to actual cases, including those that would appear on television or in newspapers; to understand it more thoroughly, I wrote academic articles about the method; to force myself to explain it as clearly as possible and to listen to the critiques of others, I introduced the method in my lectures to clinicians and students. In short, I immersed myself in the four quadrants.

WHC BIOETHICS CONSULTATION INTAKE FORM

Date: 12 Aug 2011   **Patient's Name:** Mrs Rebecca Jones   **Patient's Room:** 4E

Attending Physician (and any other Physicians involved in the case):

Dr Tom Bingham

Medical Facts: Check all that apply

ESRD	DIALYSIS	VENT/TRACH	COPD	CAD/CHD/CVD	CA	DM	CVA	PEG	ANOXIC EVENT
X	X	✓	✓	✓	✓	✓	X (bilateral)	✓	X

Other relevant medical facts:

Very poor prognosis. Irreversible brain ischaemia. On haemodialysis. History of hypertension.

Social Worker involved in case?

X Yes   ❑No

Consult called by: Dr Tom Bingham

Bioethics Consult team member(s): Dr Alex Nathanson

Ethical issues perceived:

Permissibility of withdrawing treatment

Actual Ethical Issues:

End-of-life options

Withdrawal of treatment

Is the patient capable of making his/her own decisions? If not, who is making decisions?

❑Yes   X **No**   ❑Unknown   Name/Pager

Does the patient have an advance directive?

❑Yes   X **No**   ❑Unknown

Family members and/or significant others involved in the case?
Adult son

Fig. 1.3 WHC bioethics consultation intake form (reproduced with permission)

ETHICS CONSULT FORM

PATIENT DEMOGRAPHICS

Patient Name: Mrs Rebecca Jones
Sex: F
Age: 44
Ethnicity: Black or African-American
Admission date: 1 August 2011
Discharge date: -
Room/Bed: 4E
Physicians involved: Dr Tom Bingham

ETHICS CONSULT

Date of ethics consult: 12 Aug 2011
Reason for consult:
DNAR status
Goals of treatment
Advance care plan
Person initiating consult: Dr Tom Bingham (physician)
Ethics staff responsible for consult: Dr Alex Nathanson

ANALYSIS

Family information: daughter
Medical facts: 44, F, known history of hypertension; bilateral CVAs, ESRD on haemo-D. Very poor prognosis. Irreversible brain ischaemia.
Ethical analysis – main issue: withdrawal of life-sustaining treatment
Ethical analysis – secondary issues: palliative care, DNAR status, withholding and withdrawing treatment, treatment plan.

Patient's wishes: unknown
Does patient have an advance directive? No
If yes, is a copy included in the patient notes? N/A
Can the patient make own healthcare decisions? No
If no, is there a substitute decision maker? No
Justification for role as substitute decision maker: N/A

RECOMMENDATIONS

Recommendation (medical)
Continued aggressive treatment is not medically indicated. Palliative care is appropriate.

Recommendation (ethical)
Withdrawal of dialysis combined with non-aggressive palliative approach is ethically appropriate.

Recommendation (legal/risk management)
N/A

Recommendation for follow-up
Ethics team
Family/heathcare team meeting

COMMENTS
Dr Nathanson spoke with Dr Bingham about possible withdrawal of dialysis and cessation of continued aggressive intervention. It was agreed that this course of action was ethically desirable, and that the patient's medical condition was such that further treatment would

not confer any benefit, medical or otherwise, to the patient. The medical team does not have an ethical obligation to provide non-beneficial treatment.

The patient's adult son has been consulted and is aware of his mother's unfavourable prognosis. After discussion of the harms and benefits of the various options, he is in agreement about the desirability of a palliative approach.

A DNAR order should be written and placed prominently in the patient's notes. The son has also been informed of his mother's DNAR status. The care plan is limited to symptom management. Palliative care has been called. Bleep Dr Nathanson (459) if needed.

Fig. 1.4   WHC ethics consult form (reproduced with permission)

Journals such as the *Journal of Clinical Ethics, Clinical Ethics* and the *Journal of Hospital Ethics* have regular case analyses, using a variety of analytic frameworks (although be aware of the differences between the US and UK context, especially legal ones). If you gather the cases in a file, and divide them into topics (such as "neonatal ethics", "paediatric ethics", etc.), you will soon possess a handy resource which can be used in research, writing, teaching and presentations.

Finally, you can arrange internships with hospital ethicists in North America or elsewhere.[22] In a busy bioethics department, you will encounter dozens of cases in a matter of days. Some institutions may even pay for the airfare or accommodation

---

[22] The Center for Ethics at Washington Hospital Center, Washington D.C., has a twice-yearly Clinical Ethics Immersion course, which aims to teach aspiring ethicists how to 'do' clinical ethics.

in exchange for a presentation or two. With a pro-active attitude, opportunities abound for exposure to clinical ethics cases.

References

Ackerman T, Strong C (1989) A casebook of medical ethics. Oxford University Press, Oxford

Adedeji S, Sokol D, Palser T, McKneally M (2009) Ethics of surgical complications. World J Surg 33(4):732–737. doi:10.1007/s00268-008-9907-z

Andre J (2002) Bioethics as practice. University of North Carolina Press, Chapel Hill

Audi R (2006) Practical reasoning and ethical decision. Routledge, New York

Bean R, Bean W (1950) William Osler: aphorisms from his bedside teachings and writing. Henry Schuman Inc., New York

Beauchamp T, Childress J (2009) Principles of biomedical ethics, 6th edn. Oxford University Press, New York

Bermingham V, Brennan C (2008) Tort law. Oxford University Press, Oxford

Caldicott Y, Pope C, Roberts C (2003) The ethics of intimate examinations–teaching tomorrow's doctors. British Medical Journal 326:97–99. doi:10.1136/bmj.326.7380.97

DeMarco J, Ford P (2006) Balancing in ethical deliberation: superior to specification and casuistry. J Med Philos 31(5):483–497. doi:10.1080/03605310600912675

Fins J (2006) A palliative ethics of care. Jones and Bartlett Publishers, Boston

Gardner M (1978) Aha!. Freeman and Company, New York

Gilbert D (2006) Stumbling on happiness. Harper Perennial, London

Gillon R (2005) Families and genetics testing: the case of Jane and Phyllis from a four-principles perspective. In: Ashcroft R, Lucassen A, Parker M, Verkerk M, Widdershoven G (eds) Case analysis in clinical ethics. Cambridge University Press, Cambridge

Hinman L (2000) Seeing wisely: learning to become wise. In: Browne W (ed) Understanding Wisdom. Templeton Foundation Press, Philadelphia

Hodgetts T, Mahoney P, Russell M, Byers M (2006) ABC to <C>ABC: redefining the military trauma paradigm. Emerg Med J 23(10):745–746. doi:10.1136/emj.2006.039610

Holmes OW (1896) The path of the law. Applewood Books, Massachusetts

Hursthouse R (1999) On virtue ethics. Oxford University Press, Oxford

Jones J, McCullough L, Richman B (2008) The ethics of surgical practice. Oxford University Press, Oxford

Jonsen A, Siegler M, Winslade W (2010) Clinical ethics, 7th edn. McGraw-Hill, New York

Kasparov G (2007) How life imitates chess. William Heinemann, London

Kuczewski M, Pinkus R (1999) An ethics casebook for hospitals. Georgetown University Press, Washington

Kushner T, Thomasma D (eds) (2001) Ward ethics. Cambridge University Press, Cambridge

Lantos J, Meadow W (2006) Neonatal bioethics. Johns Hopkins University Press, Baltimore

Lo B (2000) Resolving ethical dilemmas, 2nd edn. Lippincott Williams & Wilkins, Philadelphia

McCullough L (2005) Practicing preventive ethics. Physician Executive 31(2):18–21

Menzel P, Dolan P, Richardson J, Olsen J (2002) The role of adaptation to disability and disease health state valuation: a preliminary normative analysis. Soc Sci Med 55(12):2149–2158. doi:10.1016/S0277-9536(01)00358-6

Mental Capacity Act (2005) c.9 http://www.legislation.gov.uk/ukpga/2005/9/contents

Ortiz D (2007) Strong magic. Kaufman and Company, Washington

Pannick D (1987) Judges. Oxford University Press, Oxford

Peterson S (2011) Caring attention in the healing relationship with the obese patient. J Hospital Ethics 2(2):6–9

Shenk D (2010) The genius in all of us. Icon Books, London

Sherman N (2010) The untold war. W.W. Norton & Company, New York

Sokol D (2007) Ethicist on the ward round. Br Med J 335:670. doi:10.1136/bmj.39344.636076.59

Sokol D (2008) Heroic treatment. Acad Med 83(12):1166–1167. doi:10.1097/01.ACM.0000341968. 56616.95

Sokol D (2009) The slipperiness of futility. British Medical Journal 338:b2222

Sokol D, McFadzean W, Dickson W, Whitaker I (2011) Life and death decisions in the acute setting; an ethical framework for clinicians. Br Med J 343:d5528

Syed M (2010) Bounce. HarperCollins, London

Wittgenstein L (1963) Philosophical Investigations 1, 2nd edn. (trans: Anscombe G). Blackwell, Oxford

Chapter 2
Clinical Ethics on Paper

It is my sincere belief that all practising clinicians have at least one idea, observation, or case that could form the basis of a medical ethics article. It is a matter only of identifying what that is, and putting it on paper in the right form.[1] Armed with the ability to identify and analyse an ethical problem, you are in a position to publish articles and to contribute to the literature. Potential outlets include general medical journals, specialty journals, medical ethics journals, and newspapers. This chapter provides guidance on how to publish in clinical ethics. Again, it is based on my own experience as an author, reviewer and editor. It does not purport to be beyond dispute.

2.1 Permission

Unless identifying details are removed, patient consent is necessary. This forms part of a clinician's duty of confidentiality. If the patient is dead or does not have capacity, a relative's permission is usually required. If the patient is a child, the parents' consent is needed, as well as the child's permission if he is sufficiently mature to understand the situation. Note that 'identifying details' does not refer only to information such as name and date of birth. The article should not allow readers to infer, by joining up the dots, the identity of the patient. A highly unusual case in a particular hospital at a particular time will raise alarm bells among editors. Ask yourself "if the patient read this, would he know that it was about him?".

Sadly, without patient consent, your fascinating and exciting case may be so stripped down that only the bare bones will remain. At times, when reliant on the specific facts of the case, you will have to abandon the idea. One option is to scrap

---

[1] "The idea or angle for an article is half the struggle", one of my graduate school professors used to say. If writing about end-of-life ethics, for example, how will your article differ from the thousands of others? Wise is he who writes about less explored but important areas of medicine. Disaster medicine and military medical ethics have been relatively neglected by ethicists and provide rich opportunities for scholarship and publications (see Appendix 2), but every medical specialty has ethical issues that have been overlooked in the literature.

D. K. Sokol, *Doing Clinical Ethics*, SpringerBriefs in Ethics 1, DOI: 10.1007/978-94-007-2783-0_2, © The Author(s) 2012

the case in favour of a hypothetical one. This resolves the confidentiality problem, but tends to have a lower impact on readers. For this reason, it is worth seeking the patient's consent whenever practicable.

On the few occasions that I have had to obtain consent, I have never had my request declined. The request for permission should explain the importance of the article and offer to send the patient a copy before publication. Keep hold of the written consent, as editors may later ask for it. Note also that some journals have their own patient consent forms.

2.2 Choosing the Destination

Before putting pen to paper, decide where you want to publish. This may seem obvious, but I have lost count of the times I have received a near-final draft asking for advice on the appropriate place to submit it. This should have been determined much earlier.

In deciding the appropriate destination, ask questions such as "what am I hoping to achieve with this article?", "where is it likely to have the biggest impact?", and "who do I want to read this?". The most prestigious, high-impact journal will not necessarily be the most fitting place. Your intended readers may be specialists, not generalists. They may be members of the public, rather than clinicians. Or the article may be UK-focused, and of no interest to an American journal.

Some editors expect authors to explain in their cover letter why they have chosen their journal. A good reason, for example, is that the journal has recently published articles on the same or similar topic, and that your article pushes the debate forward. Other editors pay little or no attention to cover letters. Dr Kamran Abbasi, editor of the *Journal of the Royal Society of Medicine* (JRSM), writes:

> Cover letters divide editors of scientific journals. Many editors ignore them, dismissing them as mere marketing. Other editors use them to achieve a quick sense of the importance of the paper and the capabilities of the authors to express their ideas. For authors, the best option is to write a short, clear, informative cover letter that can be easily adapted in case of rejection. A good cover letter has three components: a summary of the key message of the paper, an attempt to quantify the importance of the work, and an explanation of how the paper is relevant to the readers of this particular journal. A sensible author will also be polite, modest, and check that the cover letter is addressed to the editor of the journal it has been sent to.[2]

Here is an example of a cover letter that accompanied a submission to the JRSM in 2007:

Dear Dr Abbasi,

William Osler and the jubjub of ethics; or how to teach medical ethics in the twenty first century

---

[2] Abbasi K, personal communication, 18 August 2011.

I have pleasure in enclosing this short essay for publication in the JRSM. The paper was originally delivered to the Osler Club of London in May 2007 and was written with the JRSM in mind. It is not a historical piece on Sir William Osler, but uses Osler's views on medical education to present a novel argument. It calls for medical ethics to be taught on the wards, rather than in the classroom. Although this idea was proposed by Dr. Mark Siegler in the United States in 1978, it has—to the best of my knowledge—never been articulated in print in the UK.

I believe the piece, if published, will help prompt a debate amongst clinical teachers and ethicists on the value of hospital-based ethics teaching.

The article is written for a UK-based clinical audience with no ethics jargon. It uses the lessons of history to explore a topical issue in medicine and medical education (how should ethics be taught to doctors?). Given the JRSM's readership and its adventurous spirit, I feel the article would be quite at home in its pages.

Yours sincerely,

Daniel Sokol

Cast an eye on the acceptance rate of the journal, and modify your expectations accordingly. It is silly to despair at rejection from the *New England Journal of Medicine*. However, a high acceptance rate should not affect the effort you devote to the article. It will bear your name. Wherever you publish, your professional reputation is at stake, and it would be regrettable to sully it by submitting a substandard article. Reputation can be lost in an instant, and take a long time to restore. At the time of submission, the identity of the reviewer is unknown. It could be your consultant, or your future boss, or the head of the Deanery. And if it somehow slips through the net and gets published, hundreds of your peers will think less of you from your association with poor research. As with their clinical work, clinicians submitting articles for publication should strive for quality, not mediocrity.

Once the target journal is selected, find out if the journal can accommodate an article on clinical ethics and, if so, under what section. Frustratingly, some specialist medical journals do not have a section appropriate for an ethics article. If not a regular reader, the easiest ways to find the answer are by looking at the journal's website (under 'Instruction for Authors') and by asking a friend who is familiar with that publication. Everyone's time is wasted if an author submits a case report to a journal that does not accept case reports.

Also important are the word limit of the relevant section and the formatting and referencing specifications (e.g., how many references, if any, you are allowed and in what style). Read the last few articles from the section to get a sense of what the section editor enjoys. If still unsure if your article is appropriate for the section, find the name of the relevant editor (on the website or by telephone) and write a carefully crafted e-mail asking for his views. I say 'carefully crafted' because that e-mail is not just an enquiry; it is also a pitch. You are aiming to pique the editor's interest and to receive a response saying "Yes, I'll take a look at it. Send it over."

2.3 Writing the Article

Once aware of the target journal's requirements, the writing process can begin. There is no single 'right' way to write an article, but it is worth drafting a structure first. This should lead to a more coherent article, with each section and paragraph leading naturally onto the next, and reduces the likelihood of significant omissions. Under the broad structure (e.g., introduction, background, methodology, analysis, etc.), jot down the main points. This document will serve as a template for the main draft.

'Academic' or 'scholarly' is not synonymous with 'boring'. Most readers will stop reading when an article is boring, and as an author you should want people to read your work. Write with energy, but maintain a formal, professional style.

A former editor of a national newspaper once told me that I should write with a hypothetical reader in mind. This reader is sitting on a foldable seat in a busy underground carriage, ready to turn the page at any point. Although the editor's advice related to newspaper writing, the principle applies to many kinds of writing, including academic writing. Think of your own reading habits. Readers of journals also suffer from a short attention span and, unless necessary for their research or examinations, they will happily skip to the next article. The process of writing is a constant struggle to keep the reader's eyes fixed on your text.

You may have heard of the 'aha' moment. It refers to that magical moment when the solution to a problem becomes clear, when ignorance gives way to understanding. "The sudden understanding or grasp of a concept is often described as an 'Aha' moment—an event that is typically rewarding and pleasurable", writes journalist Rick Nauert, "usually, the insights remain in our memory as lasting impressions (Nauert 2011)." The monkey's ingenious solution, recounted in the previous chapter, of climbing on the researcher's shoulders to grab the banana is an example of an 'aha' moment.

In all your articles, you should strive to trigger at least one 'aha' moment in the reader. You should know what it is before you start writing. It may be PARQ or some other helpful acronym, an ethics checklist tailored to your specialty, a suggestion that will change practice, or even an observation or story so fascinating that it puts a smile on the reader's face. That insight should cause the reader to think, at the end of the article: "that was definitely worth reading".[3] A graph of the reader's interest from start to finish should look like this (Fig. 2.1):

Filled with expectation, the interest level is relatively high at the start of the reading process, wanes naturally after a few paragraphs, rises sharply with the 'aha' moment, dips again after the high, and rises with a strong finish. Stray below 1, and the reader will move on to the next article.

Maintaining the reader's interest requires you to know the readership. An insight for surgeons may be a platitude for radiologists. If you remember only

---

[3] Appendix 7 contains a short article in which I aimed to include a number of 'aha' moments in quick succession.

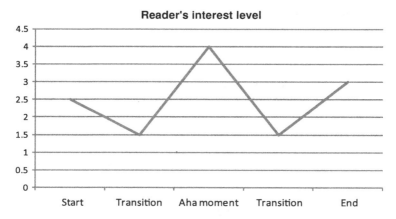

Fig. 2.1 Interest level of a hypothetical reader

one thing from this chapter, remember the importance of the 'aha' moment. It will boost your acceptance rate.

If writing about an individual case, try to derive general lessons. It will give the article broader significance, and answer the all-important "so what?" question. What does this case teach us about good medical practice? What was done well, and what can be improved? What effect, if any, has this case had on clinical practice at your institution?

A journal article does not need to be written in the style of a PhD. It is generally junior academics who, in a bid to appear scholarly, seek to impress with fancy words. The masters value clarity of language. Remove all superfluous words and sentences. If jargon is unavoidable, give a brief definition unless confident that the readers of the journal will understand it. Again, this is impossible without knowing your readership. Neurologists will be familiar with Dandy-Walker syndrome, a congenital brain malformation, but it will be Greek to most general practitioners. Similarly, spell out acronyms in full on first use. Use active, not passive sentences: "we conducted an ethical analysis" is preferable to "an ethical analysis was conducted".

A final tip on the writing itself, from an editor's perspective: avoid spelling errors and typos like the plague. They create a distinctly unfavourable impression. I once reviewed a paper which contained a typo in the first word. Read and re-read the article until you are quite sure there are no errors. Be your own, ruthless editor. Once you have focused on the micro-level of the word and sentence, zoom out to the level of the paragraph. Make sure each one flows naturally into the next. Then send the article to a friend or colleague for a fresh pair of eyes.

Remember to acknowledge that person at the end of your article, but ask them for permission first. If your informal reviewer makes significant suggestions on the content, which you later adopt, consider adding him as an author.

2.4 A Word on Authorship

Consult the authorship criteria for the journal. As hard as it may be, do not include anyone who does not satisfy the criteria. Not even your consultant. Acknowledge them at the end if they have helped you. That is the purpose of the 'Acknowledgements' section. Adding 'gift' authors dupes the reader, gives the bogus author a false appearance of expertise, and devalues the contribution of the real authors. It breaches your professional code, namely the obligation to be honest and trustworthy, and it has been known to backfire on the bogus author. If the research proves to be fraudulent or in some other way unethical, the bogus author is left in a difficult situation.[4]

Similarly, do not leave out anyone who fulfils the authorship criteria. That is also deceptive. If a person has done enough to be an author, add him.

Authorship criteria

Many medical journals subscribe to the criteria of the International Committee of Medical Journal Editors (ICMJE). An author must satisfy all three of the criteria below:

1. Substantial contribution to conception and design, acquisition of data, *or* analysis and interpretation of data;
2. Drafting of article *or* revising it critically for important intellectual content;
3. Final approval of the version to be published.

Note that, under the ICMJE guidance, obtaining funding for the research, collecting data, or supervising the research group are not, in themselves, sufficient to constitute authorship.

If there are multiple authors, try to agree the final authorship order early on. This helps avoid later disputes between authors.

2.5 Invitations to Resubmit

It is rare to receive an outright acceptance. Most of the time, you will be asked to make changes. Bitterness and anger directed at the reviewers are common responses. Yet, do not reveal any trace of disappointment in your response. Thank the reviewers for the opportunity to improve the article, and point out your modifications. Something like this is fine:

---

[4] An eminent professor of obstetrics and gynaecology found himself in hot water, and in the pages of tabloid newspapers, when he added his name to a publication written by one of his team (Jaffer and Cameron 2006). The lead author had fabricated the data. When the truth came out, the professor was forced to resign from the presidency of the Royal College of Obstetrics and Gynaecology (RCOG) and from the editorship of the RCOG journal.

Dear [name of editor],

Many thanks for giving us the opportunity to revise our manuscript. As requested, we have addressed the reviewers' comments. These are detailed in the paragraphs below.

Then explain how you have addressed *each* of the reviewers' comments. If you disagree with a reviewer, say so diplomatically and explain why. The more detailed your response, the better, but do not waffle. A comprehensive cover letter will show that you have taken the reviewers' comments with the seriousness they deserve. If the editor decides to send the manuscript back to the reviewers, they are likely to be impressed by the thoroughness of your response.

I have learnt the importance of good revisions the hard way. I submitted an editorial to a leading medical journal soon after I finished my Master's in medical ethics. It was one of my first submissions. After peer-review, the editor asked me to resubmit with changes. I looked at the reviews, spent 10 min on the revision, making only the easiest changes, and fired back a slightly modified manuscript. I did not bother writing a cover letter. The article was rejected. When I recounted my disappointment to a more experienced friend, he was flabbergasted at how little effort I had put into the revision. An invitation to resubmit is only a short step from acceptance, so avoid the temptation to cut corners. Since that experience, my cover letters have been meticulously detailed.[5]

2.6 Rejections

Even eminent authors get rejected, although the more eminent you become the more you will be invited to write articles, by-passing some of the hurdles. Until you attain that status, do not be disheartened by rejections. A rejection can lead to a better article.

If the rejection is accompanied with comments or reviews explaining why it was rejected, make appropriate changes before submitting to another journal. If the reviews were reasonably positive, you may wish to include them in the cover letter, accompanied by details of how you have modified the article. The editor will appreciate your honesty, and may expedite the review process. The fastest acceptance I ever received was in a submission to a specialty journal. The piece had been rejected from a general medical journal a week earlier with fairly positive reviews. I included them in the cover letter to the new journal, along with a detailed explanation of the changes. The unconditional acceptance landed in my inbox 15 min after pressing the 'submit' button.

---

[5] One clear and methodical approach is to address each reviewer's comment as follows:

Reviewer's comment 1: (insert the comments here either *verbatim* or in summary.)

Response 1: (include your response to the comment. Avoid the temptation to dismiss the comment as idiotic or to ignore it altogether. The editor may decide to send the document to the reviewer in question.)

Modification 1: (include the specific modification to your article here.)

Then continue with **Reviewer's comment 2**, **Response 2**, **Modification 2** and so on.

If the rejection comes without any reviews, think hard before submitting it to another journal of the same type. The original journal must have rejected it outright for a reason. Re-read the article carefully and find ways to improve it. Consider recruiting another author to help identify and correct the probable weaknesses in the paper.

Remember always to change the formatting and referencing to match the requirements of the new journal. To do otherwise suggests rejection from one journal and immediate, unaltered submission to another. It smacks of laziness or desperation.

2.7 Writing for Medical Ethics Journals

There are important differences between writing for medical journals and specialist medical ethics publications, such as the *Journal of Medical Ethics*, *Clinical Ethics*, *Bioethics*, and the *Cambridge Quarterly of Healthcare Ethics*. There are also considerable differences within medical ethics journals, so the general rule of reading the guidelines for authors applies whenever you submit to a different ethics journal.

One advantage of writing for medical ethics journals is that they tend to have higher acceptance rates than the general medical journals or the higher impact specialist medical journals. It is generally easier to survive the initial cull and reach the review stage. This means you will often receive helpful feedback, even if the article is ultimately rejected.

A possible disadvantage is that virtually all medical ethics journals have relatively low impact factors. Further, the readership is generally smaller than for medical journals. If the purpose of your article is to effect a change in clinical practice, or your target audience is junior doctors, then submitting to a medical ethics journal is a poor choice. However, if your aim is to stimulate thought, prompt a debate, and establish yourself in the field of medical ethics, then it is ideal.

Note that medical ethics journals tend to accept longer articles than medical journals, allowing authors to develop arguments more fully. The abstracts are usually unstructured, and should be short and to the point. Remember that, along with the title, the abstract is the most visible part of the article to readers, reviewers and editors, so do not rush it.

To illustrate, here is an abstract from an article I co-authored with Dr Josip Car, published in the *Journal of Medical Ethics* in 2006, which looked at the issue of telephone consultations (Sokol and Car 2006). Although brief, it conveys in broad terms the problem, its significance, and our proposed solution. In my view, an abstract is also a pitch, or an attempt to hook the reader in, so the abstract reveals the 'aha' moment (the idea of a password system to protect confidentiality), hoping that this will cause the reader to read on. Such an abstract would be inappropriate for a mainstream medical journal, but is fine for a medical ethics journal:

Abstract

Although telephone consultations are widely used in the delivery of healthcare, they are vulnerable to breaches of patient confidentiality. Current guidelines on telephone consultations do not address adequately the issue of confidentiality. In this paper, we propose a solution to the problem: a password system to control access to patient information. Authorised persons will be offered the option of selecting a password which they will use to validate their request for information over the telephone. This simple yet stringent method of access control should improve security while allowing the continuing evolution of telephone consultations.

As the articles can be longer and the readership is more versed in ethics, ethics journals generally expect greater ethical content than medical journals. You may want to use the four principles or the four quadrants to examine a case or an issue, conducting a full analysis and exploring opposing arguments. You may have space to discuss the wider relevance of a case or an issue, and make links with the existing literature on medical ethics. While it would be appropriate to devote several paragraphs introducing the four principles in a medical journal, a few lines would be quite enough in a medical ethics journal. Words such as 'deontology' and 'utilitarianism' would not need definitions.[6]

A higher word limit does not mean the article must reach that length. A common complaint among editors is the excessive length of many submissions. Cut out words. Ruthlessly.

2.8 Get an Ethicist on Board

Inviting an ethicist to help you early on can avoid some of the pitfalls of writing an ethics article. You may have to explain the nature of the phenomenon you are writing about, but overall it will probably save you time. Professional ethicists will know about the journals, their scope, their readership, and some of the recent or

---

[6] For readers unfamiliar with those terms:

Utilitarianism is a type of *consequentialist* moral theory. Consequentialists believe an act is morally right or wrong based only on its consequences. What are good consequences? For classical utilitarianism, the ultimate good is pleasure or happiness, so the consequences of an act should be measured in terms of the amount of pleasure or happiness produced by the act. In short, for a classical utilitarian, morality is about maximising happiness and minimising unhappiness.

Deontology, unlike consequentialism, places duties (the 'deon' in deontology comes from the Greek word for 'duty') and rights at the centre of ethics. For a deontologist, morality cannot be reduced *merely* to consequences. Note that in France medical ethics is called 'déontologie médicale'. For a succinct and lucid account of some key ethical theories, I recommend Piers Benn's *Ethics* (Benn 1998).

age-old debates in the field. They may notice important books or articles missing from your paper, and can include background information and details that will add an extra dimension to your work.

Ethicists can also remove tell-tale signs that you are still learning the language of bioethics. Just as doctors can generally tell if a medical ethicist is not a doctor, ethicists can generally spot when the author is not a professional ethicist. For some readers and editors, an ethicist will inject a dose of legitimacy. A clinician and ethicist combining forces to write on clinical ethics form a strong team, at least on paper, while a clinician-only or ethicist-only team may raise eyebrows among purists ("what do clinicians know about ethics? What do ethicists know about the nitty–gritty of clinical practice?").

Medical schools are a good hunting ground in the search for an ethicist. If fortunate enough to find several of them, aim for the Teaching Fellows, Lecturers or Senior Lecturers. They are most likely to need publications for their career advancement, and hence to collaborate. Check their webpage to see if their interests include clinical ethics, and if they have published in any of your target journals. Most ethicists I know would welcome a joint project with a clinician, as long as they do not feel exploited. There must be mutual benefit. Write them a polite e-mail, explaining the project and its importance to practice, and inviting them to collaborate.

If you expect your article to contain more than a minimal amount of law, it may be worth getting a lawyer on board. The law is ever-changing, and there may have been recent developments in the legislation or in the common law. Lawyers are generally harder to find, but your hospital or medical school may have a legally qualified person who can check a draft or join as a co-author.

2.9 Writing for Newspapers

If you are after a wide readership, or trying to inform the public, newspapers are a natural target. Someone once said that the average academic article has five readers.[7] Newspapers, or websites such as BBC Online, can have hundreds of thousands, although the readership is diffuse and non-specialist. The shelf life of the article is also short. It will appear in the print version of the newspaper for one day and the online version for a few more before disappearing in the recesses of the virtual universe. A major advantage is the instant feedback from readers, but brace yourself for negative comments.

In this age of electronic publishing and lightly moderated online responses, the point about negative comments also holds true for academic publications, although

---

[7] I once spent two years of my life forging a satisfactory definition of the word 'deception' and published it in a respected bioethics journal. Six months after publication, I had received no comments on the article. A year later, no comments. Two years later, no comments. I doubt five people have read that paper.

they will usually not be as vitriolic as newspaper responses. If your article is even slightly contentious, you may find yourself at the receiving end of critical remarks, some personal. Do not take them at heart. There are strange folk out there. Use your judgement to decide whether to post a response. You may wish to wait a few days and address several comments at once in your response. *Always* be courteous and professional, however rude the initial respondent.

The same advice on knowing the publication and its readership applies to newspaper writing. In what section will it fit? What is the word limit? In newspaper articles, the opening paragraph is crucially important. Many readers do not read past the first paragraph, or even the first line. It is common for editors, whose inbox may be full of unsolicited submissions, to reject articles based on the first paragraph. It is, for me, the longest to write. The second longest is the last paragraph. Aim for a strong start and a memorable finish. In between, break up the text with regular paragraphs written in simple, engaging prose. Picture your reader on that foldable seat in the underground train.

All the health editors I know are happy to hear from doctors and other health professionals. Clinicians can provide insights that other journalists cannot. Newspaper editors have a *penchant* for real-life cases, so by all means include them, but remember to respect patient confidentiality. Newspapers do not have the same checks as journals, and no publication is worth a visit to a disciplinary hearing. As with medical journals, ask yourself "would the patient recognise himself when reading this article?".

A major advantage of submitting to a newspaper is the quick verdict. If you get a rejection, submit it to another editor. Submitting the same article to several editors at the same time—called 'multiple submissions'—is tempting, but frowned upon by editors. Do not do it. If no response is forthcoming, send a short reminder e-mail to the editor, stressing the urgency.

If you send your pitch to the wrong person, you are unlikely to receive a response. Find out the name of the relevant editor. It will usually be the health editor. If not listed online, contact the switchboard and ask for the name and e-mail address of the health editor.

Once you have established a rapport with the editor, it will be easier to publish for that newspaper. They may even commission pieces. If so, ask about the pay and the deadline. Respect that deadline at all costs, especially if it is a daily newspaper.

2.10 The Pitch

Before submitting your completed article, send a pitch to the relevant editor, detailing your idea. If you have written for newspapers before, mention this. If you have not, make the pitch as punchy and persuasive as possible. Explain who you are, and why you are in a position to write the piece. Set out the idea in one or two paragraphs, giving an idea of the content and stressing the main point of the article.

Richard Warry, an editor at the BBC News website, advises to 'make the piece relevant to the widest possible audience'.[8] Remember that editors think in terms of headlines. Try to include a case or anecdote, and if relevant attach a graph, table or image to accompany the piece.

Below is an actual example of a successful pitch to a newspaper editor. As I had written for him before, there is no personal introduction.

From: Daniel Sokol
Sent: 01 November 20XX
To: [ ]
Subject: New piece for Health
Dear [ ],
I hope you're well.
Recently, a GP told me that one of his patients, an old lady, believed that doctors were bound by "that Oath" never to tell patients the truth. Yesterday, talking to some doctors over dinner, I realised that many doctors are as clueless as the old lady. Although everyone's heard of the Hippocratic Oath, very few people know anything about it.
The proposed piece would be a very short guide to the Oath, which should help dispel the many myths about it. I also will also link it to present day medicine. I think it could be of interest to your readers, both medical and non-medical. What do you think?

Your piece may have a better chance of getting accepted if there is a 'peg', a recent item in the news that makes your story topical. If the article is already written, wait for a peg and act quickly once it appears. Contact the editor immediately with the idea.

If you decide to submit a finished piece, make sure that it is the appropriate length for the section and send it along with your pitch. Say that you are willing to make changes.

A word about payment. The rates will vary from about £100 to £1,000 (for some tabloids) for a 1,000-word article. There is some room for haggling, but not much. Avoid haggling for the first few submissions, as you may acquire a reputation as a difficult customer.

Even respected broadsheets make so-called 'adjustments', not all of which are favourable. A few years ago, I wrote a piece on what used to be called the 'killing season', the time in August when newly qualified doctors start their first job. After interviewing medical educators and doctors from all levels of the hospital hierarchy, I concluded that medical students are adequately trained and that patients have nothing to fear from a visit to hospital in August. A few days later, the article appeared in print with the headline 'Danger, white coats—be very afraid, says Daniel Sokol'. When I expressed my displeasure to the editor, her reply was brief: "sorry, but there was no story otherwise".

---

[8] Warry R, personal communication, 8 July 2011.

2.11   Submitting an Ethics Abstract at a Medical Conference

As well as publications, proficiency in medical ethics can lead to oral presentations at conferences. If you have taken the trouble to write an article on ethics, consider giving a presentation based on it. Similarly, if you have presented on an ethical topic at a meeting or conference, consider turning it into a publication.[9]

The difficulty in submitting an abstract on ethics is the unnatural fit between the sections in the website's abstract submission page, which are tailored to the average medical study, and your ethics project. Still, you must play the game. Below is an example of an ethics abstract submitted to an international neurosurgical conference. It was accepted as an oral presentation.

Introduction

Junior neurosurgeons regularly perform operations at the limit of their competence. Although often supervised, their operative proficiency may not match those of their more senior counterparts. This may result in longer operating times, a higher incidence of errors, and an increased risk of morbidity to the patient. An apparent tension exists between the need to train neurosurgeons and the duty to act in the patient's best interests. Is it ethically permissible to subject patients to trainee surgeons who may not achieve the best results? If so, what if anything should the patient be told about the operating surgeon?

Methods

The ethical issues will be analysed using the four principles of medical ethics described by Beauchamp and Childress.
The principles of beneficence, non-maleficence, respect for autonomy, and justice, will be applied systematically to identify and analyse the ethical dilemmas arising from the practice of allowing neurosurgical trainees to 'train' on patients.

Conclusion

There are compelling arguments in favour of allowing trainees to operate on patients, based on a broader interpretation of beneficence and non-maleficence which encompasses both present and future patients. However, a more open approach to informed consent may be required to comply with the demands of respect for patient autonomy.

---

[9] Be aware, however, that some journals refuse submissions which include data already presented at conferences, or which appear in an abstract.

2.12 Presenting on Clinical Ethics at Meetings and Conferences

If you succeed in publishing in journals or newspapers and impress at conferences, you may well receive invitations to speak at study days, departmental seminars, grand rounds, and even as a plenary speaker at major conferences. There are not many medics with a specialist interest in ethics, nor ethicists with a specialist interest in clinical ethics.[10]

For the presentation itself, structure is just as important as in an article. Present the facts, identify the problems, use an analytic framework, draw your conclusions, and end strong. Time yourself so that you do not exceed your limit, and leave the allocated time for questions. A common mistake when clinicians present on ethics is to spend too much time on the clinical facts and not enough on the analysis. The analysis consequently appears thin or rushed, and this gives the unfortunate impression that the speaker know precious little about ethics.

Anticipate questions from the audience, and prepare good answers. While it is OK, or even desirable, to know the wording of a few key sections, do not under any circumstances read out a pre-prepared text of the entire talk. This is the kiss of death of any lecture. If you are not comfortable presenting, consider courses or books on presentation skills. This will prove valuable for the rest of your professional, and personal, life. It is also relevant to the subject of the next chapter: teaching medical ethics.

References

Benn P (1998) Ethics. Routledge, London
International Committee of Medical Journal Editors (2011) Uniform Requirements for Manuscripts Submitted to Biomedical Journals: Ethical Considerations in the Conduct and Reporting of Research: Authorship and Contributorship. http://www.icmje.org/ethical_1author.html. Accessed 3 July 2011

---

[10] Yet, do not expect invitations to come pouring in after only one or two publications, even if they appeared in 'heavyweight' journals. In his autobiography, Steve Martin described his elation at appearing on the renowned American TV programme *The Tonight Show*. He expected instant recognition.

Here are the facts. The first time you do the show, nothing. The second time you do the show, nothing. The sixth time you do the show, someone might come up to you and say, '"Hi, I think we met at Harry's Christmas party." The tenth time you do the show, you could conceivably be remembered as being seen somewhere on television. The twelfth time you do the show, you might hear, "Oh, I know you. You're that guy." (Martin 2007, pp. 125–126).

In my experience, the same is true in medical ethics. It takes time to get known, or at least sufficiently known that conference organisers and journal editors think of you when deciding who to invite or commission.

Jaffer U, Cameron A (2006) Deceit and fraud in medical research. Int J Surg 4(2):122–126. doi:10.1016/j.ijsu.2006.02.004

Martin S (2007) Born standing up. Pocket Books, London

Nauert R (2011) Insights on the 'Aha' moment. Psychcentral.com. http://psychcentral.com/news/2011/04/01/insights-on-the-aha-moment/24906.html. Accessed 16 July 2011

Sokol D, Car J (2006) Patient confidentiality and telephone consultations; time for a password. J Med Ethics 32:688–689. doi:10.1136/jme.2005.01441

Chapter 3
Teaching Clinical Ethics to Medical Students and Clinicians

An excellent way to learn more about clinical ethics is to teach it.[1] Teaching encourages you to read the literature and to understand it well. It gives you an opportunity to explain concepts and answer questions in a less threatening environment than a conference. It can help boost your confidence as an ethicist. And, importantly, it can be fun and rewarding.

A great lecture is:

- Informative or thought-provoking
- Relevant
- Clear
- Non-threatening
- Entertaining

While these characteristics may appear self-evident, mediocre lectures are regrettably common. Trainee barristers have weekly teaching sessions on how to speak in court, each video-recorded for careful and critical evaluation, but after our PhDs I and other junior academics were unleashed onto students with only the most rudimentary training in presentation skills. Thankfully, like all skills, teaching can be learnt, and even the most bashful can become excellent teachers.

Similar advice applies to teaching ethics as to writing articles. Know your audience. If the likely constitution of the audience is unclear, ask the person who invited you. How much clinical experience do the audience members have? First year medical students will generally have little knowledge of, say, intensive care, so explain the context before discussing the issues. Explain why the subject is important.

When asked to teach specialists, find out the sort of ethical issues they encounter. Plastic surgeons are unlikely to care much about the ethical issues faced

---

[1] Clinicians wishing to gain some teaching experience should contact the person in charge of medical ethics at their local medical school, highlighting their clinical experience and their personal interest in the subject, and offering to conduct a session either alone or in conjunction with another teacher.

by pharmacists. Call up a friend who works in that speciality and ask for details, then follow up with a search of the literature to see what, if anything, has been written on the issues.

If you do not know a specialist, search the literature on Pubmed, Google, or individual journals. Sometimes, there is scant literature, or the existing literature is old, or from another country. It is sensible to ask the person who invited you for examples of recent ethical problems encountered by their unit. You can then use those as a 'peg' on which to hang your material. The audience will be impressed by your familiarity with the issues, and the content will resonate with them.

Even if you have given a similar talk before, tailor it to the audience. A talk on whether doctors should always tell patients the truth should differ from a talk on whether *nurses* should do so. Nurses face different challenges, and have their own code of ethics. They are generally in greater contact with patients, and have additional duties towards the doctor in charge of the patient. The examples used should involve nurses, not just doctors. To do otherwise risks alienating much of the audience.

As with writing, display boundless enthusiasm. Do not allow the audience to be bored. The comedian Steve Martin said that teaching is a form of show business (Martin 2007, p. 86). All your preparation, all your wisdom, all your wonderful material, will be lost if the audience members are not awake, or if they are listening to their ipods instead of you. Use realistic cases, preferably authentic ones in which you have been involved.

For example, when teaching about the challenges of obtaining informed consent,[2] I state that it is not uncommon for surgical patients to be 'consented' on the morning of a major operation. I then describe a real case, recounted to me by a colleague who works in a prestigious hospital in the United States. A porter was

---

[2] Appendix 8 contains an article in which I describe some of those challenges.

Fig. 3.1 Teaching slide on the principle of respect for autonomy

taking a female patient to the operating theatre. During the journey, the patient expressed her relief at having the operation. "I'm so glad I'll be able to get on with my life after this", she said, "and finally start a family". She was about to have a hysterectomy. The porter stopped the trolley, called the medical team, and the operation was postponed.[3] This case illustrates the fact that some patients, even moments before the operation, are ill-informed about the procedure and its consequences.

Use photos, videos and props to create and maintain interest. Look out for scenes in TV and film that can demonstrate a point. Intersperse your delivery with discussion, or the showing of a video clip, or some other activity that breaks the monotony. As with your hypothetical readers, assume your audience has a short attention span.

It is currently *à la mode* to split students into groups, where they can discuss a problem amongst themselves. If that works for you, fine. I have always preferred opening up the floor for discussion. Splitting people into groups can interrupt the flow of the talk and infantilise the audience. A large group has its drawbacks, but it allows greater control of the group, and of the direction of discussion.

PowerPoint slides are recommended, unless the talk is short or your presentation skills first-rate. Judiciously used, slides can create a more engaging experience for the audience. Use text sparingly, and photos often. Do not write in full sentences, but use bullet points. If intending to spend a few minutes on, say, respect for autonomy, write only a few key words. This forces the audience to listen. A sample slide on respect for autonomy, used in my own teaching, appears below (Fig. 3.1):

The slide has 13 words, giving a basic definition of the principle ("deliberated self-rule") and four key duties derived from the principle. The photo, which provides some visual interest, shows an operation in a government hospital in southern India. It serves both to anchor the abstract concept of respect

---

[3] This case is recounted briefly in the article in Appendix 9. The article also provides ideas on what to teach junior doctors about medical ethics and law.

for autonomy in clinical reality, and as a springboard for a personal anecdote. If teaching medical students, the accompanying 'patter' might go as follows (commentary on the patter is in *italics*):

The first principle I want to look at may be familiar to some of you (*minor suspense. Transition to slide above to reveal principle*). Who has heard of the principle of respect for autonomy? (*involve audience. Show that you are willing to interact*) Yes, quite a few of you (*acknowledge response, display spontaneity*). 'Autonomy' can be understood as 'deliberated self-rule' (*a touch obscure, so explain more clearly*). The principle of respect for autonomy, in the context of medicine, refers to the obligation to respect patients's choices based on their own beliefs and values. How, then, can doctors fulfil this obligation to respect the autonomy of their patients? (*At this point, I would elaborate on the four duties listed on the slide, explaining the relationship with the broader principle of respect for autonomy*)

Now, although respect for autonomy is a strong principle here in the UK, it is not so everywhere in the world. I found this out for myself a few years ago (*introduce idea of cultural variations in bioethics, shift from theory to real life*).

I spent a month with a surgeon in South India (*point to photo*). The first operation I saw was an above-knee amputation on an elderly diabetic patient. He had a gangrenous foot and needed the operation to survive. The day after the operation, the surgeon asked about my interests. I answered "medical ethics, especially truth-telling in medicine". "Ah", the surgeon replied, "that is most interesting (*brief pause*). You remember the old patient yesterday?". "Of course". "Well", he said, "I did not tell him that I was going to amputate his leg. I tricked him into having the operation". (*gasps of horror from the audience*).

In this country, such behaviour is unthinkable. Respect for patient autonomy prohibits it. The guidance of the GMC prohibits it, and the law considers it a crime. But for this surgeon, working in a remote village in India, respect for autonomy was not the most important principle. In the surgeon's mind, it was trumped by a stronger obligation: to save the patient's life (*transition to next slide on principle of beneficence*)

Magicians have a basic rule: leave the audience wanting more. The same applies to teaching. When your lecture ends, the audience should be slightly disappointed that it has ended so soon. If you can achieve this, repeat invitations will pour in. And remember, as with articles, to finish strongly. This may require you to memorise the words accompanying the final slide, although the delivery should always appear natural. Ironically, it takes practice to appear natural.

The slide below (Fig. 3.2) completes my talk on truth-telling in the doctor-patient relationship, in which I argue that benignly intended deception may, in rare circumstances, be ethically justified.

Note the slide's simple layout and the use of the photograph.[4] Before displaying the slide, I inform the audience that the presentation is coming to an end. This

---

[4] The photo of H.L. Mencken is in the public domain (the copyright has expired) and so can be lawfully used.

Fig. 3.2 Final slide of truth-telling talk

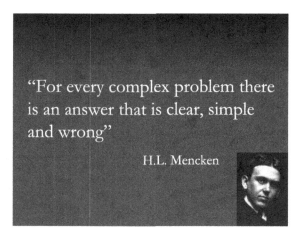

raises the attention level. "I end this lecture with a short quote from the American journalist H.L. Mencken". I then read the quote, which generally elicits some smiles and laughter. I always end with the same lines: "The issue of truth-telling in medicine is complex, which is why I think so many clinicians and philosophers have struggled with it for centuries. I believe we will continue to struggle with the issue for centuries to come (*pause*). But the more I think about the truth-telling problem, the more I'm convinced that the absolutist position that clinicians should always (*emphasis on 'always'*) tell patients the truth and never, ever deceive their patients is clear (*short pause*), simple (*short pause*) and wrong (*short pause*). Thank you". As well as signalling the end of the presentation, the 'Thank you' also acts as an applause queue.

Make your audience feel at ease. This requires *you* to be comfortable. An audience can sense a nervous speaker, and members may feel uneasy as a result. They may label you an amateur, and take you less seriously. If you are not confident, feign it. At the risk of stating the obvious, stand up straight, speak up, do not fiddle with pens, paper or other objects, and maintain regular eye contact with the audience, including those at the back and the sides of the room (Fig. 3.3).[5]

Do not take yourself too seriously. Laugh and smile when appropriate. Encourage those who ask questions, and always adopt a charitable interpretation. If you do not know an answer to a question, say so with confidence. If you do not make a big deal out of it, neither will the audience. No one can know everything.

If asked to teach medical students, find out if you are expected to cover part of the curriculum. If so, ask if there is an existing presentation or handout on the subject that you can inspect or modify. There is no need to follow the template slavishly, however, so inject your own personality and anecdotes into it.

---

[5] There is no better way to improve your presentation skills than to film yourself giving a lecture and reviewing the performance afterwards. It will probably make for uncomfortable but valuable viewing.

Fig. 3.3 The author at the 10th World Congress of Bioethics, Singapore, maintaining eye contact with the audience members in the far left corner of the room

Find out the anticipated size of the audience in advance. It will help your preparation by giving you an idea of how many handouts you need and how interactive you can be with the audience. Will you be in a small room, where it is easy to break students into groups (if that is your thing), or will you be in a lecture theatre with fixed seats? Knowing the size of the audience will affect the nature of your relationship with the audience. It is harder to establish a rapport with a large group. You will also need to project your voice more in a big room. In short, you will be more at ease when you arrive at the venue if you know how many people to expect.

Ask the organiser if the audience has heard lectures on a similar topic. If so, refer to it at the beginning of your talk: "I know that one of my colleagues talked to you last month about the ethics of end-of-life care. Today, I want to focus on one particular area which he may have mentioned in his talk: advance decisions."

If planning to use visual aids, check if the facilities are available. Send a copy of the slides to the organisers in advance, so that they can install it on the computer. Take a copy of the talk on two separate USB sticks on the day, in case one malfunctions or is rejected by the local computer.

Finally, a practical suggestion. If you intend to give talks regularly, buy a slide-changer. It is a small device that allows you to change slides remotely, and serves also as a laser pointer. No longer stuck behind the laptop, or condemned to press 'return' on the keyboard after each slide, you can position yourself in the most suitable spot. Some even have an integrated timer. It really is a wonderful gadget.

While the prospect of teaching ethics can be daunting, especially teaching practitioners who may not be receptive to the subject, keep in mind the Chinese proverb: 'enjoy life, it's later than you think'. So, as you step out onto the stage, remember to enjoy the experience.

Reference

Martin S (2007) Born standing up. Pocket Books, London

Chapter 4
Submitting an Application to a Research Ethics Committee (REC)

Applying for research ethics approval is an exercise that many clinicians will have to conduct at least once in their career. In the UK, obtaining REC approval for clinical research is a legal obligation, and many journals will require you to confirm the approval prior to publication.

The application process fills some with dread, and stories of extraordinary delay and injustice abound. Last month, a medical registrar recounted to me how, after a grilling by members of the REC, her consultant swiftly left the room and cried. Yet, with proper preparation, there is nothing to fear. REC members, a motley crew of pharmacists, statisticians, researchers and interested lay members, seldom bite, and then only if provoked. As for delays, RECs must deliver their verdict within 60 days of receiving a *valid* application. Often, it is much faster.

Avoid thinking of the REC as an enemy, intent on sinking your project. They exist both to protect patients and to help researchers. They may identify issues that you have overlooked and even suggest ways to improve the project from a scientific or methodological perspective. It is more helpful to view the REC as a benevolent but strict mentor.

4.1 Audit, Service Evaluation, or Research?

The type of project determines if ethics approval is necessary. The table below (Table 4.1), drawn from the website of the National Research Ethics Service (NRES 2011), helps distinguish between audit, service evaluation, or research. NRES is the body which regulates all RECs in the United Kingdom, and its website provides excellent guidance on all stages of the process (http://www.nres. npsa.nhs.uk/).

Sometimes, the table will not resolve the question. In such cases, do not assume that ethics approval is not needed. Contact the Chair or Administrator of the local REC for advice.

D. K. Sokol, *Doing Clinical Ethics*, SpringerBriefs in Ethics 1,
DOI: 10.1007/978-94-007-2783-0_4, © The Author(s) 2012

Table 4.1 Differences between research, audit, and service evaluation (reprinted with permission)

Research	Service evaluation	Clinical audit
The attempt to derive generalisable new knowledge including studies that aim to generate hypotheses as well as studies that aim to test them.	Designed and conducted solely to define or judge current care.	Designed and conducted to produce information to inform delivery of best care.
Quantitative research—designed to test a hypothesis. Qualitative research—identifies/explores themes following established methodology.	Designed to answer the question: "What standard does this service achieve?"	Designed to answer the question: "Does this service reach a predetermined standard?"
Addresses clearly defined questions, aims and objectives.	Measures current service without reference to a standard.	Measures against a standard.
Quantitative research -may involve evaluating or comparing interventions, particularly new ones. Qualitative research—usually involves studying how interventions and relationships are experienced.	Involves an intervention in use ONLY (The choice of treatment is that of the clinician and patient according to guidance, professional standards and/or patient preference).	Involves an intervention in use ONLY (The choice of treatment is that of the clinician and patient according to guidance, professional standards and/or patient preference).
Usually involves collecting data that are additional to those for routine care but may include data collected routinely. May involve treatments, samples or investigations additional to routine care.	Usually involves analysis of existing data but may include administration of simple interview or questionnaire.	Usually involves analysis of existing data but may include administration of simple interview or questionnaire.
Quantitative research—study design may involve allocating patients to intervention groups. Qualitative research uses a clearly defined sampling framework underpinned by conceptual or theoretical justifications.	No allocation to intervention: the health care professional and patient have chosen intervention before service evaluation	No allocation to intervention: the health care professional and patient have chosen intervention before audit
May involve randomisation	No randomisation	No randomisation
Requires REC review	*Does not require REC review*	*Does not require REC review*

Even if a project does not require REC review, it may still raise ethical issues. These must be addressed. Aside from the professional obligation to conduct research ethically, a reputable journal may refuse to publish an article containing ethically dubious research. If the ethical violation is serious, editors can contact your institution or, if appropriate, the General Medical Council.

4.2 Build a Multi-Disciplinary Team

An ethicist can help improve your research ethics application, and may have prior experience of submitting such forms. He may even sit on a REC. In contrast to invitations to co-author an article, an invitation to read through an application form is hardly an appealing prospect to an ethicist. Offer something in return, such as inviting the ethicist to form part of the research team and to collaborate on resulting publications.

If your research involves statistical analysis, and unless you are very comfortable with statistics, it is wise to seek the advice of a professional statistician. Every REC has at least one statistician, and many an applicant has floundered when questioned about study design and methodology. In the words of one statistician REC member, "nothing worries me more than an application where the Principal Investigator claims to be an expert in statistics, but it is obvious from their CV that they are no such thing". A multi-disciplinary approach to the research generally results in a smoother ride through the application process.

4.3 Completing the Form

All RECs will contain lay members and experts from fields quite different to your own. You should have them in mind when completing the application form. Do not assume much knowledge of the clinical environment, and explain drugs, procedures, and other technical terms. For this reason, cutting and pasting from a protocol or other document is not recommended. Tailor the text to your audience.

Obscure passages in the application form are frustrating because they require multiple readings and visits to Google. A REC member might spend two hours carefully reading the form, and still only have the faintest idea about the project. Lack of clarity can lead to misunderstandings about the project. Write in full, active, and short sentences, free from spelling errors and typos. Remember that most RECs do not pay their members. Make the life of the committee members as easy as possible.

Thoroughness is a virtue when completing the section on ethical issues. More than any other, that section reveals the ethical sophistication or otherwise of the applicant. The ethical ignoramus, lacking in moral perception, will write that there are no ethical issues when a moment's reflection suggests otherwise. "It is a questionnaire study", they might write, "and hence does not raise any ethical issues". This invites disaster. Even the simplest questionnaire can raise issues of recruitment, confidentiality and data protection. The ethically astute applicant will write down all the relevant issues, anticipating any risks of harm, even remote ones, and their likely severity and probability.[1]

Harm is not restricted to physical harm, but can include anxiety or embarrassment. When volunteering as a hospital magician many years ago, I decided one day to conduct some research of my own. I wanted to experiment with a new magic trick. I went to the oncology ward, and found a side-room occupied by a middle-aged woman, clearly bored with her Sudoku puzzles. I explained that I was the hospital magician, and asked if she would welcome the opportunity to win £50. Her boredom vanished at once. She gave an enthusiastic nod. I showed her five envelopes and slipped a £50 note into one of them. I then shuffled the envelopes and numbered them one to five. "You have four chances", I told her. "If you pick the envelope with the money, you can keep it". Three tries later, two envelopes remained on the table. One of them contained the money. "You've had three goes, and have only one more; that's a 50% chance of winning £50. Choose carefully". I must admit that my own heart was racing, as there was a distinct possibility of error on my part. As a student, I could hardly afford to lose £50. It was with considerable relief, then, that I tipped out the £50 note from the only envelope the woman had not selected. When I looked up at her face, tears welled up in her eyes. She was distraught. "How can you take advantage of vulnerable people like that?", she muttered.

Although obvious now, it had not occurred to me that my experiment could cause any harm. I had wrongly assumed that my 'volunteer' was the same as a guest at a wedding or a diner at a restaurant. In my excitement, I had forgotten that this particular 'volunteer' was suffering from cancer and potentially emotionally vulnerable, or that £50 may have seemed a vast sum of money to her, or that the disappointment of losing could itself constitute harm. The story reveals a failure of moral perception on my part, and highlights the importance of thinking broadly about the potential harms of a proposed intervention, even one which may at first sight appear innocuous.[2]

4.4 The Patient Information Sheet (PIS)

The PIS is a common source of unhappiness among REC members, and readers are advised to read the guidance issued by NRES on information sheets, available on the NRES website (Hunter 2011).

---

[1] Some research, of course, raises no material ethical issues. An example is research on existing tissue samples where consent to do so has already been obtained. In such cases, it may be possible for researchers in the UK to expedite the application process through 'proportionate review', where a sub-committee of the REC will review the proposal. Consult the NRES website for further details: http://www.nres.npsa.nhs.uk/applications/proportionate-review/.

[2] Several readers of a draft of this book have asked if, in an act of redemption, I offered the woman the £50. I did not.

A recurring complaint, at least in my REC, relates to the readability of the document. It should be written in language suitable for the anticipated readership. A PIS for patients waiting at a GP surgery should read quite differently to a PIS for astronauts. The PIS for the first group should reflect the fact that the reading age of some patients may be very low. The PIS must give a clear idea of what the research is about and what participants will be expected to do. It must be user friendly, and brief. The average patient will be baffled by terms such as 'plantar fascia exercise' or 'proprioception'. Avoid complex or specialised terminology, unless you are confident that *all* the prospective participants will understand them. Use the active voice and short sentences.

Include subsections to break up the text or, as my committee recommends, use a 'question and answer' format. The questions may include:

- What is the purpose of the research?
- Who is doing this research?
- Why have I been invited to take part?
- Do I have to take part?
- What will I be asked to do?
- What is the device or procedure that is being tested?
- What are the benefits of taking part?
- What are the possible disadvantages and risks of taking part?
- Can I withdraw from the research and what will happen if I don't want to carry on?
- Will I receive any expenses or payments?
- Will my taking part affect the medical treatment I am receiving?
- Whom do I contact if I have any questions or a complaint?
- What happens if I suffer any harm?
- What will happen to any samples I give?
- Will my records be kept confidential?
- Who is organising and funding the research?
- Who has reviewed the study?

Avoid formulations which sound threatening, such as 'You are expressly forbidden from drinking tea or coffee during the study', and do not overload the patient with information. It will confuse rather than clarify. A lot of the advice above is common sense but it is surprising how often it is ignored.

4.5 The Meeting

Once the application is submitted, the REC will send a letter inviting you to attend the meeting. Move heaven and earth to make it in person. Without a good explanation for the absence, many committees hold a dim view of researchers who fail to attend. Committee members are likely to have questions about the project and, if you are not there to answer them, they may either reject the application or invite you to the next meeting in one or two months' time.

If you have a project supervisor, ask him to attend too. I have seen junior doctors left helpless in REC meetings, unable to answer questions, sent by their supervisor like a lamb to the slaughter. The supervisor, or principal investigator, should generally be present.

As well as arriving on time and dressing smartly, remember to take a copy of the application form and all relevant documentation (e.g., questionnaires). Members may refer you to specific sections of your application, and you will look silly if you do not have it at hand. I remember clearly the rebuke of a barrister REC member aimed at an applicant who had not heeded this advice: "Do you think I can turn up in court and tell the judge "sorry, I don't have the papers with me?" I'd be laughed out of court".

If, despite your attentive efforts in completing the form, you need to make amendments after the initial submission but before the meeting, inform the chair or administrator of those changes and send the latest version. The new version can be passed on to the members, who will have time to read and evaluate it. If you introduce new material at the meeting, be prepared for grumblings of discontent. Members do not like to read material for the first time at the meeting itself, under time pressure.

Prepare for the meeting carefully. Anticipate likely questions, and good answers to them: "Will volunteers feel pressured if invited to participate by the medical team?",[3] "What if you discover something potentially harmful from the blood tests?",[4] "Is there a risk that the sensitive questions will cause distress to the participant?",[5] "Why are you not reimbursing travel costs?",[6] "Why are you

---

[3] Ideally, the person recruiting the participants should not be involved in their medical care. If this is not possible or practicable, explain why and describe the ways in which any pressure will be minimised.

[4] It is common for researchers to promise strict confidentiality when, in fact, they are willing to breach it if there is serious risk to the participant or others, or if there is an admission of serious professional misconduct or criminal behaviour. It is conceivable, for example, that in an interview-based study on how surgeons prepare themselves psychologically for major operations, or pilots before a long-haul flight, a participant may reveal that he drinks alcohol to calm his nerves. Would guaranteeing absolute confidentiality be appropriate in such cases? It is important, therefore, to anticipate possible revelations which may put participants or others in danger. Needless to say, it is improper to promise strict confidentiality if this cannot be guaranteed. Participants must be told if there are circumstances in which their confidentiality will be breached.

[5] Questions on mental health, childhood and other potentially sensitive issues can awaken unpleasant memories in participants and cause emotional distress. It is advisable to show the REC that you are aware of this possibility and, if appropriate, to offer participants the contact details of an appropriate counselling service.

[6] It is generally unreasonable to expect participants to bear the costs of travelling to the venue if the sole purpose of the visit is to take part in the research.

offering £50 for participation?",[7] "How much time will the volunteers have to decide on participation?",[8] and so on.

Do not ignore a risk if one exists. At times, you will have no choice but to acknowledge the risk, explaining either that the risk is small or that you have taken all reasonable means to minimise it as far as possible ("Ah, finally, a reasonable applicant!", the committee will think to itself).

REC members do not wish to impede your research. They want the research to be conducted to high ethical standards. As this is also your aim, enter the room with a spirit of cooperation rather than confrontation. The meeting is an opportunity to talk to interested people about your research.

Always be civil, even if the questions appear daft or misguided. Do not, as some applicants do, call the chairperson by their first name. Some members of the committee may consider this discourteous. Address them as 'chairman' or 'madam chairman'. If you sense the committee has misunderstood an important aspect of the research, politely clarify the situation. If possible, raise the issue during the meeting itself, not afterwards. If you do not know the answer to a question, be honest. RECs are skilled at detecting moonshine. If appropriate, offer to find the answer after the meeting and to communicate it promptly to the chair. At the end of the meeting, thank the committee members for their time and comments (even if your heart is not in it).

When you receive the decision letter, expect some revisions and recommendations. The majority of applications receive conditional approvals. The changes required are usually minor, and should take little time to incorporate. Further, the revisions may improve your study.

If interested in the workings of a REC, contact your local REC and ask to observe a meeting. Even a single meeting should help you distinguish a good application from a bad one. You will applaud the well-prepared researchers, cringe at the unprepared ones, and promise yourself never to belong to the latter group.

References

National Research Ethics Service (2011) http://www.nres.npsa.nhs.uk/. Accessed 2 Aug 2011
Hunter D (2011) A hands-on guide on obtaining research ethics approval. Postgrad Med J. Online first doi:10.1136/pgmj.2010.109348

---

[7] There is nothing unethical in itself with offering money to cover participants' time and expenses. What must be avoided is offering such a substantial sum of money that potential participants may feel unduly influenced. You must therefore be prepared to justify why the payment is reasonable in the circumstances. Another commonly ignored issue with payment is how much, if anything, the participant who withdraws early from the study will receive.

[8] Ideally, researchers should give potential participants at least 24 h to decide on participation. If that is not possible, you must be prepared to justify the shorter period to the REC.

Chapter 5
Conclusion and Appendices

The world of clinical ethics is brimming with opportunities. It is a young discipline, and its future success lies in part on the active involvement of enthusiastic clinicians. There is much that clinicians can do to promote ethics in their institution or specialty. They could set up a consultation service or an ethics committee in their hospital, pilot an ethics checklist in their unit or conduct ethics-related research, organise a talk or a one-day conference on the ethics of the specialty, submit abstracts on ethical topics in conferences, articles to journals, or chapters to medical textbooks, or just raise ethical issues on ward rounds and team meetings. It is a field for pioneers.

Readers eager to deepen their knowledge of medical ethics should consider courses on the subject. These range from short but intensive five-day courses to Master's degrees which take one year full-time or two years part-time. An Internet search should reveal the courses in your area. The search may not indicate if the course has a clinical or philosophical focus, so ask the course director for details. You can also enquire about the background of past delegates (are they mostly doctors, nurses, or non-medics?) and the professional background of the faculty. Attending a short course before committing to a Master's programme is prudent, although I declare my competing interest as co-director of a short course.

Whatever your purpose for picking up this book, be it to learn how to analyse an ethics case or publish on medical ethics, I hope it has demystified the process of 'doing ethics'. Now that you have studied the chart map, as Osler would say, it is time to set sail. If you have the time and the will, do send news of your voyage.

D. K. Sokol, *Doing Clinical Ethics*, SpringerBriefs in Ethics 1,
DOI: 10.1007/978-94-007-2783-0_5, © The Author(s) 2012

5.1   Appendix 1: The Dilemma of Authorship

As a graduate student in the humanities I remember being surprised at the tales of bogus authorship recounted by my counterparts in the sciences. One person would do virtually all the work, another would give useful feedback, another would glance at the final version, while yet another would be just someone who worked in the same department—and all would be co-authors of the published manuscript. "It happens all the time," the scientists would say. I nevertheless ascribed such practices to a pocket of ambitious, amoral scientists in the cut throat environment of a major research institution.[1]

With time I discovered that this was not at all unusual in science and indeed in other disciplines. In the months leading up to the UK Research Assessment Exercise, whose outcome determines a department's academic reputation and share of government funding, I heard of academic ethicists adding the names of struggling colleagues to their publications. Thus I cannot but look on multi-authored publications with suspicion, despite the authorship criteria and other strategies adopted by many academic journals with fine intentions.

Recently a young surgeon approached me with a "tricky situation." Earlier that day a more senior surgeon had asked to be a co-author of his now completed paper. He had not contributed in any way to the project but needed the publication for career reasons. The other surgeon's consultant had advised him to piggyback on the junior colleague's work. The awkwardness arose because the other surgeon now asking to be a co-author had been most helpful in training that young surgeon

---

[1] From Sokol D (2010) The dilemma of authorship, British Medical Journal, 336:478.

in the operating theatre. "He's been so nice to me," he remarked, "but he hasn't done a thing related to this paper." Furthermore, the surgical team worked well together, and the surgeon did not want to sour relations in the firm by turning down the request and upsetting his colleague and the consultant. "And does it really make a difference?" he continued. "I won't compete with him for jobs, and I'll still be first author, right?" What advice would you give this troubled surgeon?

It will come as no surprise that I suggested he politely refuse, explaining that the journal requires him to sign a form listing authorship criteria, which in his case would not be met. The junior surgeon could also tell his colleague that, although this particular project is complete, he would be delighted to work with him on another paper. I did not advise him to give a detailed justification for the decision, unless asked for it by the colleague. Invoking words such as honesty, trust, fairness, professionalism, and academic integrity would only highlight the inappropriateness of the initial request, make the colleague feel morally attacked, and sound obnoxiously self righteous.

The eagle eyed among you will have noted that the formulation "the journal requires me to sign a form" could imply that, were it not for that wretched form, the young surgeon would be happy to grant co-authorship. If the phrasing is disingenuous, this cannot be more than a moral peccadillo. If it is morally wrong, it is trivially so. The twin goals of declining a request for undeserved co-authorship and maintaining good relations with a kind colleague take priority and require skilful diplomacy.

But was this advice, however tentative, too demanding? By placing so much moral weight on the requirements of justice and lofty principles, did I evince an insensitivity to the practical realities of the situation and the hierarchical structure of the surgical team? Did I overlook the surgeon's self regarding duties of preservation? Team harmony and personal relationships are important considerations. Personal disputes at work create an unpleasant environment and may lead, through poor communication or low morale, to poorer care of patients. Without the help of his senior colleague, the young doctor's clinical skills may not develop as rapidly. And of course these are anxious times for doctors seeking scarce training posts—all the more so for surgeons. Rightly or wrongly, applicants are turned down for lack of peer reviewed publications. To risk irritating a senior colleague who has regular contact with a consultant who writes references is, in the current climate, a dangerous game to play.

Although I feel strongly that this lamentable situation needs to change, I struggle to see workable solutions to the problem. I do not even know whether I gave the surgeon sound advice. Words of William Osler seem pertinent here: "I have tried to indicate some of the ideals which you may reasonably cherish. No matter though they are paradoxical in comparison with the ordinary conditions in which you work, they will have, if encouraged, an ennobling influence."

As an ethicist I draw comfort from the surgeon's moral unease at the request. Less reflective colleagues may not have perceived it as an ethical issue at all. It is sad, however, that he should even be confronted with this moral dilemma. To claim authorship in an article to which one has made no contribution is to perpetrate a fraud on the reader. It is incompatible with the ideals of authenticity and honesty espoused by the profession.

Despite the indisputable nature of these ideals, the practical task of changing bogus authorship is a daunting one, requiring a change in mentality across the medical hierarchy, from old school consultants to newly minted doctors.

5.2 Appendix 2: The Medical Ethics of the Battlefield

Athena, goddess of war, gave Asclepius two vials of the Medusa's blood. The blood from Medusa's left side could raise the dead; the blood from her right side could kill instantly. The descendants of Asclepius—the thousands of medics who today grace the battlefields of the world—rarely use the right sided blood. Battlefield euthanasia, in which death is hastened to avoid prolonged suffering, is a controversial practice; but it is as old as war itself and, whatever laws or rules prohibit it, will doubtless continue until wars cease. In this column, however, I wish to focus on the dilemmas associated with the left sided blood. When should it be used and when forgone? And who should benefit from it?[2]

The ability to maintain the wounded alive is nothing less than astounding. Medical advances, combined with improved body armour and rapid evacuation, have resulted in lives saved that would have been unsalvageable only 20 years ago. A recent visit to Headley Court, the Defence Medical Rehabilitation Centre, brought home to me the remarkable recoveries of soldiers who, weeks before, were lying on the battlefield on the brink of death. Yet, as in the civilian setting, the power to revive the dying has brought with it a host of ethical difficulties.

In one scenario, a member of the local Afghan security forces has suffered massive injuries from an improvised explosive device. He has lost both his legs and both his forearms. The blast has removed his entire face. Tourniquets are controlling the bleeding from the legs. He is still alive. If he can be saved by use of the coalition forces' state of the art medical services, what of his future once he is transferred to a local health centre, whose facilities pale in comparison?

One Canadian paramedic working in Kandahar, Afghanistan, in 2007 described the transfer of patients to the local hospital as a "death sentence" (Kondro 2007:134). The hospital had no ventilators, resuscitation equipment, laryngoscope, or monitoring devices. Kevin Patterson, a Canadian doctor also posted to Afghanistan, recalls a mass casualty incident involving a mixture of coalition personnel and Afghans (Patterson 2007). The doctors were told not to intubate any of the Afghans with burns exceeding 50%. Without a burns unit, those patients would be doomed. The coalition patients, on the other hand, could be repatriated to their home countries to obtain high quality burn care. Such divergent treatment is hard to bear and highlights the need to develop local healthcare infrastructure, but what are the immediate alternatives?

Athena's vials are exhaustible, and resources problems can also plague the military medic. Beds, staff, and stocks are limited. Our patient might

---

[2] From Sokol D (2011) The medical ethics of the battlefield 343:d3877.

singlehandedly drain the hospital's blood bank, leaving nothing in reserve for future casualties. The third revision of the US Department of Defence's manual *Emergency War Surgery* states that "the decision to commit scarce resources cannot be based on the current tactical/medical/logistical situation alone" (US Department of Defence 2004). Such decisions should be made with an eye to the future.

If our Afghan patient is treated and survives to discharge, what kind of life awaits him back in his village, where the realities of survival and attitudes to profound disability may be a far cry from our own? This question cannot be answered without an understanding of the local culture, religion, and outlook. It is morally dangerous to uniformly impose our interpretation of when it is desirable to live or die, dismissing the patient's views as backward, barbaric, or misguided.

If the decision to treat is made, the patient will need to be evacuated. A medical emergency response team (MERT) helicopter can arrive within minutes to provide advance life support and whisk our patient off to intensive care at a state of the art "role 3" medical facility. Yet, there is another consideration. Every excursion by the MERT carries risk. The helicopter is vulnerable and prone to enemy ground fire, and this additional danger must be factored into the decision.

There is another factor, relevant in this context but seldom encountered in civilian medical ethics: morale. Dwight Eisenhower called morale the "greatest single factor in successful wars" (Charlton 1990:144). Allowing the soldier to die on the battlefield can damage the morale of the troops. It smacks of abandonment. The fact that the patient is Afghan provides an added reason to evacuate him, for not doing so may cause other Afghans to lose faith in the commitment of their fighting partners.

In October 2010 the Defence Medical Services organised a day long meeting to discuss some of the ethical issues facing medical personnel in the field, including scenarios such as the one set out in this column. This was a significant step, a recognition that pre-deployment training should include an appreciation of the ethical challenges that can otherwise startle the unwary medic. When Athena gave Asclepius the vials, she did not provide advice on their use. The Defence Medical Services are working to fill that gap. I cannot remember the last time I left a conference with so many unanswered questions swirling in my mind.

The literature in military medical ethics is growing but still pitifully small. My hope is that experts from relevant fields will devote more attention to one of the most challenging, important, and fascinating areas of medical ethics.

5.3 Appendix 3: Ethicist on the Ward Round

Not so long ago in the *BMJ* I quipped that most professional medical ethicists could not distinguish their "gluteus maximus from their lateral epicondyle" and suggested that such ethicists should undergo a short clinical attachment (Sokol 2006).[3]

---

[3] From Sokol D (2007) Ethicist on the ward round, British Medical Journal 335:670.

Soon after publication, a nephrologist kindly invited me to observe a ward round at his hospital. It proved to be a puzzling experience, not because the blood gases, creatinine levels, diagnostic tests, and myriad statistics recited by a junior doctor sounded like one of Mallarmé's incomprehensible poems, but because, as the afternoon progressed, I noticed the patient-as-person fading behind this shroud of science. I felt comfortable with my consultant, my team with their dangling stethoscopes, the all-knowing computer wheeled by the bedside, and the timid patient, dwarfed by our confident crowd. Ethics seemed a million miles away.

This absence of ethics was most puzzling of all. I spend my days thinking, teaching, and writing about medical ethics, but there, in a group of doctors and with the patient before me, the subject seemed alien. "Think," I urged myself, "what are the ethical issues here?"

My reverie would soon be interrupted: "Urine output… raised creatinine levels… metabolic acidosis… abdominal x ray." Even in cases that I knew had obvious ethical dimensions, such as those involving futility and end-of-life decisions, I felt powerless to use ethical reasoning since I could not perceive the ethical issues with any clarity. It reminded me of a time when, intent on discovering a card magician's method for a trick, I got so engrossed in his patter, in Sam Spade and the evil kings (a dramatic reference to the ace of spades and the four kings), that I forgot to observe the subtle movements of the conjurer's hands and body. Magicians, like doctors, are well aware that language can disguise reality, distracting the mind from the disappointing truth ahead, be it a palmed card or a grim prognosis.

My proximity to the patient, instead of highlighting the ethical components, obscured them. The incantation of scientific jargon, the outward confidence of the consultant and his team, the austere clinical environment, and the meekness of the patient all combined to give an air of certainty to the ritual. Ethics, this antithesis of science, had no place in this assured display. I could now see why some doctors and medical students found it so hard to appreciate the relevance of ethics to clinical practice. "Ethics and medicine are inseparable," we tell our students, but up close the link is not so obvious. It may be easy enough to identify ethical issues in the classroom, but at a crowded bedside the task takes on added complexity and requires practice.

More recently, I attempted to fill the gaping holes in my medical knowledge by spending five weeks in a southern Indian hospital, observing the work of a rural surgeon. Again, I initially struggled to perceive the ethical elements. I was enthralled by the medicine, the ritual of surgery, the mesh, the corkscrew, and other instruments, the different kinds of suture material, the mattress and subcuticular stitches, the smells and sounds and techniques. But as the days went by, as I saw more surgeries, it became easier. I learnt to zoom out of the medical and focus on the social and ethical dimensions. These more uncertain, fuzzy elements of the healing endeavour began to emerge from the mass of clinical information.

As my ethical gaze slowly sharpened, I reflected on the surgeon's kind hearted paternalism and the submissiveness of patients; the considerable influence of relatives in decision making; the prevalence of disclosures that were "economical with the truth"; the limited importance of confidentiality in this communal setting; the perfunctory nature of obtaining consent; the ethical implications of treating illiterate and medically unsophisticated patients; the financial and emotional costs of surgery to poor families; the responsibilities of sleep deprived surgeons and anaesthetists towards their patients, their colleagues, and themselves; the difference a few words of comfort can make in times of pre-operative fear; the role of humour and camaraderie in the theatre; the wisdom of using mobile phones when operating; the extreme difficulty of speaking your mind when offence may result; the proper relationship between culture and ethical norms; and many other issues that were initially as invisible to me as the card magician's sleights. I was not merely thinking about clinical ethics, but actually "doing ethics," in real time with flesh-and-blood patients.

The first step to moral action is moral perception, since an ethical problem can seldom be resolved if not first spotted. For teachers of medical ethics, developing this skill in students should be a priority and the most critical place to do so is at the bedside. Suturing an orange in a lab and suturing a uterus in a casesarean section are quite different activities. The same holds true with studying ethics in the lecture hall and "doing ethics" on the wards. The aseptic first is a poor approximation of the messy second.

5.4   Appendix 4: The Slipperiness of Futility

He was shot in the back. The surgeons could not save him. He lay in bed, unconscious, his life ebbing away as blood trickled down tubes to large jars at the base of his bed. As cardiopulmonary resuscitation would have been futile, we wrote a "Do not attempt resuscitation" order. The case reminded me of the etymology of the word "futile." "Futilis" in Latin means "leaky." The patient was leaking blood from various wounds, and nothing could stop it.[4]

At a recent examiners' meeting, a professor of surgery admitted that he would have got the ethics question wrong. The question concerned the definition of futility. "So how would you define futility?" I asked. He paused and, like Humpty Dumpty in *Through the Looking Glass*, answered: "Something is futile if I say it is." This remark highlights the semantic slipperiness and subjectivity of the term "futile." Yet, in the clinical frontline, futility, coated with a veneer of objectivity, is often used as a moral trump card, a dismissive pronouncement to end all discussion: "I'm sorry. We're stopping aggressive care. It's futile."

Psychiatrists must sigh in frustration when asked whether a patient has capacity. The capacity to decide what? Similarly futility is not free floating but linked to a specific goal. Prescribing antibiotics for a viral illness is physiologically futile, but if your goal is to leave the surgery in time for the first aria in *Don Giovanni* then it is not (although this would still be a breach of your duty of care). Futility, then, is goal specific, and when you next hear colleagues say that such and such is futile you can surprise them and ask, "Futile with respect to what?"

---

[4] From Sokol D (2009) The slipperiness of futility, British Medical Journal 338:b2222.

When teaching this subject to medical students I shuffle a pack of playing cards, select a card at random, and ask whether it is futile for them to guess the identity of the card. Some say yes, others say no, and once in a blue moon a statistically minded student will ask if the two jokers are included in the pack. Never is there unanimous agreement. The point of the exercise is to illustrate the variability of our quantitative assessment of futility. Some scholars have suggested that an intervention is futile if it has not worked in the last 100 cases (Fins 2006, Schneiderman et al. 1990). Under that definition, guessing the card would not be quantitatively futile. Even if we accept this somewhat arbitrary "last 100 times" rule, in practice the problem is that it is rarely possible to know whether an intervention has worked the last 100 times, especially as no two cases are identical.

The students who believe in the futility of naming the card still venture a guess if tempted by a £50 cash prize. The perceived futility of the exercise does not translate into a refusal to try. The reason is that there is no cost associated with the guess. The benefit is potentially significant and the cost minimal. As Kite and Wilkinson point out, sometimes the reason why clinicians withhold or withdraw an intervention is not because it probably won't fulfil its purpose but because it will cause harm or deprive others of benefit. An intervention can be simultaneously futile, harmful, and wasteful (Kite and Wilkinson 2002).

One of the saddest cases I have seen involved a woman so viciously mauled by dogs that she was left in a vegetative state. When considering her resuscitation status, one of the doctors stated that, on the grounds of futility, she should not be resuscitated. When probed further, it emerged that the doctor believed that the patient's quality of life was so awful that cardiopulmonary resuscitation was not medically indicated. This is another type of futility: qualitative futility (Jonsen et al. 2006:29). It is based on a subjective evaluation of whether the goal of the intervention is worthwhile.

Although ethically aware clinicians need not be familiar with the vast literature on the concept of futility, they might wish to remember the following four points (Burns and Truog 2007, David 2008, Grossman and Angelos 2009):

- Futility is goal specific.
- Physiological futility is when the proposed intervention cannot physiologically achieve the desired effect. It is the most objective type of futility judgment.
- Quantitative futility is when the proposed intervention is highly unlikely to achieve the desired effect.
- Qualitative futility is when the proposed intervention, if successful, will probably produce such a poor outcome that it is deemed best not to attempt it.

When using the term, clinicians may be referring to several types of futility—for example, that an intervention is highly unlikely to achieve the goal (quantitative futility) and also that the goal itself is undesirable (qualitative futility). As futility is so rhetorically powerful and semantically fuzzy, doctors may find it helpful to distinguish between physiological, quantitative, and qualitative futility. This classification reveals that a call of futility, far from being objective, can be coloured by the values of the person making the call. Like "best interests," "futility" exudes a confident air of objectivity while concealing value judgments (Gillon 1997, Sokol 2008).

I conclude on a practical note. Clinicians should be wary of using the word "futile" in front of patients and relatives. As Jonsen, Siegler, and Winslade propose, it may be better to think in terms of proportionality or "the imbalance of expected benefits over burdens imposed by continued interventions" (Jonsen et al. 2006:33). Furthermore, "futile" suggests that nothing can be done. Recall the ancient medical wisdom: "To cure, sometimes. To relieve, often. To comfort, always" (Russell 2000). There is always something to be done.

5.5 Appendix 5: Heroic Treatment: Reflections on Harm

There is an amusing scene in the television series Scrubs in which J.D., a cheerful hospital doctor, gathers his interns in a huddle at the start of a day's work. "Hippocratic Oath on three," he orders, "one, two, three…." In unison, hands atop hands, they exclaim, "first do no harm!"[5]

This expression, or its Latin equivalent primum non nocere, is found neither in the famous oath nor in the Hippocratic corpus. The phrase, coined by Thomas Inman, dates from 1860, around the time of this lithograph (Inman 1860). The lithograph, depicting some unfortunate and clearly petrified patient, takes us back to a time when doctors, however benevolent in intent, often caused more harm than good (Wootton 2006). James Simpson, an esteemed professor of surgery at Edinburgh in the mid-19th century, believed surgical patients in hospitals were "exposed to more chances of death than was the English soldier on the field of Waterloo (Porter 1997:369)."

In 1850, a French physician, J. Dupuy, defended his doctoral thesis on limb amputation. He counted all amputations performed in a four-year period in his Bordeaux hospital and noted 94 amputations, 47 deaths, and a mortality rate of 50% (Dupuy 1850).

Although buzzing with the advent of modern anesthesia (1846), which along with numbing pain allowed more time to operate, these were still the dark days before Lister and his antiseptic technique (Lister had a mortality rate of 45% for major amputations in Glasgow during 1864–1865; it dropped to 15% during 1867–1869 following the introduction of his antiseptic routine) (Kirkup 2007).

With such high risks, primum non nocere was sage advice. The phrase, however, needs to be refined. Each time we attempt to benefit someone, in medicine or everyday life, we also risk harming them. We cook a sumptuous meal for friends, only to give them gastroenteritis, or utter a comforting comment to a depressed friend only to redouble their anxiety. Thus, any clinician who interprets

---

[5] From Sokol D (2008) Heroic treatment: reflections on harm, Academic Medicine 83(12): 1166-67.

primum non nocere literally ought to leave medicine, as benefiting patients often requires the infliction, or at least the risk, of harm. The surgeon cuts open the abdomen (harm) to remove the inflamed appendix (benefit). Ethicists thus talk of the obligation to avoid causing net harm. One translation might be primum non in ultimum nocere ("first, cause no ultimate harm"), but ultimum also implies "lasting harm," which is not accurate as some procedures are beneficial overall despite causing permanent damage. Hence, a neurosurgeon may excise a glioma, saving the patient's life, but at the cost of slight and permanently reduced motor function.

More precise, though less pretty, would be primum non plus nocere quam succurrere ("above all, do not harm more than succor"). I somehow doubt J.D. and his interns would bellow such a phrase. The lithograph's caption [the lithograph (not reproduced here) shows a traumatised patient strapped to a chair with no arms and legs, and reads *"This is what I looked like after what doctors call heroic treatment"*] suggests that clinicians at the time were inclined to overtreat patients. Doubtless this was true of some, yet Dupuy's thesis reveals a clear appreciation of the seriousness of amputations, and of the need to balance the risks and benefits. He observes, "it is indeed a quite sudden transition which, in a matter of hours, deprives a man of an entire limb (Dupuy 1850)."

The issue of overtreatment is also pertinent in the early 21st century. I remember a meeting in a major Canadian hospital, in which a senior clinician read an interminable list of procedures performed on a recently deceased cancer patient. When he finally got to the end, he shook his head and said, "It's not easy to die in this hospital." With ever-improving technologies and the corresponding ability to keep people alive, however dreadful their injuries and grim their quality of life, the question, "when should we stop aggressive care?" will be increasingly posed.

When patients have capacity, a reliable way to ensure that a treatment's benefits outweigh the harms is to ask them directly, giving them accurate information about the alternatives, since what we value and how we balance different values vary amongst individuals. However, this approach cannot be applied when the patient is not autonomous. Advance directives, which allow us to know the autonomous wishes of now incompetent patients, and appointed proxy decision-makers, will become even more important as new tools and knowledge keep death at bay for longer and in more situations.

At all times, we should be guided by what is best for the patient. While this may sound trite, the observation about the difficulty of dying in a state of the art hospital suggests that on occasion we treat aggressively because we can rather than because we should. This lithograph captures the horror of surgery at a time when mortality rates were sky high. It also coincides with a momentous development in medical thought: the realization in the community that medicine helped little and often caused more harm than good.

In my medical school, we sometimes ask prospective medical students at interview what they believe is the greatest advance in medicine in the last 150 years. This aforementioned realization, though an ideological rather than a technological or pharmacological breakthrough, would give antibiotics, vaccination, or imaging a run for its money. Although printed over a century and a half ago, the lithograph also prompts us to reflect on, and question, our current practices. Are we really doing more good than harm, and, if harm is inevitable, how can we benefit our patients with minimum harm? These are questions that, unlike the coats and cravats of the surgeons, will remain in fashion.

5.6 Appendix 6: The Moment of Truth

Edmund Pellegrino, a professor of medicine and a giant of medical ethics, once remarked that, for the clinician, the "moment of truth may come at three in the morning, when no one is watching." This prompted me to ponder on "the moment of truth." What is it? And can we prepare for it?[6]

The moment of truth is a bullfighting term. The "hora de verdad" refers to the moment when the matador entices the bull with the "muleta" (the red cape draped over a stick) and, with the precision of the anaesthetist hitting the epidural space in an obese patient, plunges the sword into the bull's neck for the kill. If he thrusts the sword at a slight angle he will sever the aorta and the bull will die in seconds. If the matador misses, his body is exposed to the sharp horns of the frenzied animal.

We encounter a moment of truth when we are put to the test, and how we respond becomes a measure of our worth. Sometimes, as in an acute emergency, the moment of truth is clear: the patient is hypoxic, oropharyngeal visibility is poor from the blood and swelling of trauma, and the tube must go in immediately. At other times, especially with patients with more chronic illness, the moment of truth is identified only retrospectively. A doctor may realise too late that he or she omitted something that could have prevented a poor outcome, such as the radiologist who realises that he or she missed a lesion on the x-ray picture.

The moment of truth can involve physical actions, as in the difficult intubation; decisions, as with the surgeon contemplating whether to operate; or attitudes to events or circumstances. William Osler wrote of being "ready for the day of sorrow and grief with the courage befitting a man." For Osler, that moment came years later with news of the death of his only son from shrapnel wounds in the first world war (Starling 2003).

The "truth" in the phrase "the moment of truth" can refer to true skill, true merit, or true strength of character. This helps us answer the question of how we can prepare for moments of truth. We can prepare by honing our technical competencies. The cardiothoracic surgeon Fyodor Uglov, famous for his technique, sutured 400 rubber gloves before performing portacaval anastomoses on patients (Lichterman 2008). Alone, at three in the morning, the well prepared trainee can insert that all important central line in the patient with a sudden onset of severe sepsis. It is this fear of encountering the moment of truth that, at least in part, explains why some junior doctors look on the night shift with dread.

---

[6] From Sokol D (2010) The moment of truth, British Medical Journal, 340:c1992.

We can work on developing our character, putting ourselves in situations in which we can learn to exercise virtues such as courage, kindness, and wisdom. This may require us to seek new experiences and step outside our comfort zone.

A turning point in my development as a medical ethicist was on hearing a song, "Moi mes souliers," by the Canadian singer Félix Leclerc. It was about a man's travels and adventures, from school to war, through fields of mud, through countless villages and streams. The final stanza, loosely translated, goes: "Heaven, my friends, is not the place for polished shoes. So if you seek forgiveness, hurry and get your shoes dirty." As I could see my own reflection in my shoes, I travelled to various hospitals around the world to get them dirty.

Alone, at three in the morning, the trainee with dirty shoes can then decide to reassure a frightened patient when it would be easy to pretend not to notice. Yet, even with the cultivation of skill and virtue, it is impossible to prepare fully for some moments of truth, those monumental ones defined by their life changing nature. Osler never recovered from his son's death and was prone, in private, to bouts of weeping.

Moments of truth reveal something fundamental about ourselves, and as such they represent an opportunity for self improvement. They are perhaps unique to humans. The charging bull cannot conceive of a moment of truth. Only the matador can experience it, thinking to himself, as he sees the saliva flying from the charging animal's mouth, "This is it." For Pellegrino and many medical ethicists who call themselves "virtue theorists," the focus of medical ethics should not be on what is the right or wrong action but on developing the character of the clinician, fostering the virtues that will help him or her cope with the "this is it" moments in the practice of medicine.

Respected clinicians on the wards and in the GP's surgery have a much greater influence in the development of virtue in students than my colleagues and I do in the classroom. It is difficult to teach courage or integrity in a packed lecture theatre. Virtues in medicine are learnt most effectively by watching and learning from clinicians who act virtuously. Osler believed that medicine should be taught on the wards (Osler 1906); so should the bulk of medical ethics, for ethical decisions in clinical medicine are made under conditions that cannot be recreated in a classroom. Repeated, realistic exposure is the key to good ethical training. After all, it is in the arena—with the cheering crowd, blustering heat, dazzling sun, swirling sand, and raging bull—that matadors learn the essence of their art.

5.7 Appendix 7: 'Make the Care of Your Patient Your First Concern'

The first rule of *Good Medical Practice*, issued by the General Medical Council, is: "Make the care of your patient your first concern" (General Medical Council 2006). With its strong Hippocratic flavour, the statement captures a fundamental truth about the practice of medicine, pointing to the sacred and timeless nature of the encounter between the healer and the sick person.[7]

Yet, however noble in spirit, the rule should be no more than a rule of thumb. Although "patient" is in the singular, few doctors have only one patient. Doctors must therefore choose how to allocate their "concern" among their many patients. It is neither possible nor desirable to treat each patient as a first concern, as some patients, usually the sickest, merit more concern than others. The principle of justice requires the doctor to determine which patient deserves the greatest attention.

In a field hospital in a conflict zone, four polytrauma patients are admitted after an explosion. One has multiple traumatic limb amputations. The others have less severe injuries but require blood transfusions. Treatment of the first victim will activate the massive transfusion protocol. Should the hospital's entire stock of blood and plasma be used on that one patient? In such a situation triage priority shifts from "treat those in greatest medical need" to "save the most number of lives." The care of your multiple amputee is, regrettably, no longer your first concern. The rule is modified as follows: "Make the care of your patient your first concern, bearing in mind your other patients and their particular needs."

At times the interests of the public outweigh the obligation owed to an individual patient. A doctor is under an obligation to inform the authorities of a patient with yellow fever, however much the patient may protest. The first concern is not so much the patient but protecting the population from infection. So the revised rule is now: "Make the care of your patient your first concern, bearing in mind your other patients and their particular needs, as well as any protective obligations to the broader community."

I have recently argued in this column that doctors' duty of care is not an absolute obligation, to be discharged however perilous the situation (Sokol 2009). In extreme circumstances—such as epidemics, where treating patients involves a high risk of infection and modest benefits to patients—doctors' obligations to their children, parents, siblings, and loved ones may take priority over the care of patients. The doctors who left their dying patients in the early outbreaks of Ebola haemorrhagic fever in Sudan and the Democratic Republic of Congo did not necessarily act unethically. The doctors and nurses who remained, many of whom lost their lives to the virus, acted beyond the call of duty. The rule now looks as follows: "Make the care of your patient your first concern, bearing in mind your other patients and their particular needs, as well as any protective obligations to the broader community and obligations you may have towards others for whom you are responsible."

Even in ordinary times, making the care of your patient your first concern seems too demanding. Your life, personal and professional, would be dominated by this overriding concern; your working day would be interminably long, your holidays pitifully short. Your relations with friends, family, and others would suffer. You would not conduct research, publish articles, attend conferences, conduct activities

---

[7] From Sokol D (2011) "Make the care of your patient your first concern", British Medical Journal, 342:d646.

that would further your career, or develop your skills to help future patients, for the rule ignores your personal ambitions and talks only of the present patient.

The "bare" rule, strictly interpreted, would also pose problems for trainees learning to perform procedures. If a junior doctor is anxious about inserting a central line or carrying out a cholecystectomy, the rule suggests that he or she must ask a senior colleague to do it, as this is probably best for that particular patient. A trainee is more likely than an experienced colleague to make a mistake or cause discomfort, even if supervised. Yet this logic is not conducive to learning and development. Although the present patient will benefit, future patients will suffer. Thus the updated rule is: "In your professional capacity as a doctor, make the care of your patient your first concern, bearing in mind your other patients, including at times future patients, and their particular needs as well as any protective obligations to the broader community, your own obligations to develop your skills and knowledge as a clinician, and obligations you may have towards others for whom you are responsible."

Finally the rule can be misused. I have heard doctors invoke the rule to justify their exaggerations to radiologists to expedite their patient's scans. Doctors in the United States have been known to deceive insurance companies to obtain treatments for their patients (Wynia et al. 2000). If the care of your patient is your first concern, this may lead you to flout other rules, including legal ones. So the final version of the rule is: "In your professional capacity as a doctor, make the care of your patient your first concern, acting within morally and legally acceptable limits and bearing in mind your other patients, including at times future patients and their particular needs as well as any protective obligations to the broader community, your own obligations to develop your skills and knowledge as a clinician, and obligations you may have towards others for whom you are responsible."

The first rule of the GMC is a profoundly important statement, but its brevity necessarily obscures the complexity of modern medical practice. Ironically, too literal a reading of the rule could lead to unethical conduct. It should be seen as a starting point, not a commandment.

5.8 Appendix 8: Informed Consent is More Than a Patient's Signature

The phone call came at an inopportune moment: the Friday lunchtime curry at the local Sri Lankan restaurant, usually an oasis of delectable peace away from the hustle and bustle of the medical school. "I've had a dreadful consent experience" were the opening words, "dreadful." The caller was a friend needing surgery to remove a submandibular gland.[8]

---

[8] From Sokol D (2009) Informed consent is more than a patient's signature, British Medical Journal, 339:b3224.

And so, as my curry lost its warmth, he proceeded to tell me about the strong pressure to tick the boxes on the consent form ("Oh, just tick them all—he's a very good surgeon," said the senior house officer (SHO)); about the SHO's evident ignorance of the procedure; about his distinct sense of being a nuisance ("It's half past six, and we usually go home at five o'clock," the SHO observed); and about his general unease at the whole experience. He ended his account by saying that he had, under stress, signed the consent form but that on reflection he had not truly given consent. After our discussion he cancelled the operation and opted to go private.

Ten years ago the lawyer Michael Jones published an article entitled, "Informed consent and other fairy stories" (Jones 1999). Since then informed consent has come under many attacks, for its conceptual fuzziness to its impracticability in real world medicine (Manson and O'Neill 2007). There are undoubtedly many barriers to obtaining valid consent. Some are real and deeply problematic, others are imaginary. In rural India doctors told me that it was pointless to explain interventions to patients as they were too medically unsophisticated to understand. As a lecturer whose job it is to explain philosophical concepts to students whom unkind colleagues might call philosophically unsophisticated, I was not convinced. Is it not part of a doctor's job to communicate medical information in a manner that is comprehensible to the patient?

Assessing a patient's competence to consent can also be a problem, notably in areas such as geriatrics, paediatrics, psychiatry, and neurology. At times it is not clear whether patients can understand relevant information, retain it long enough to make a decision, weigh up the pros and cons, and communicate their decision. In some parts of the country, such as east London, where many patients do not speak English, obtaining high quality consent is an ideal whose attainment is a constant struggle, all the more so if the patient speaks an uncommon language.

And what of the patient whose culture dictates that decisions be taken by the family rather than the patient? How can we reconcile this focus on the family with our atomistic notion of informed consent and respect for individual autonomy? I remember meeting an interpreter who had been asked by relatives to misinterpret the clinician so as to protect their loved one from a grim truth that would never be revealed in their home country.

Another barrier to valid consent is the skewed presentation of information. It is quite easy, through verbal and psychological manipulation, to persuade a patient to agree to an intervention (Sokol 2008). This can be deliberate or unintentional. Because of our belief in a procedure's value, or out of a concern not to worry the patient unduly, we can paint a rosy picture of the situation. At times, when a senior clinician delegates the task of obtaining consent to a junior member of the team, the junior may feel some pressure to secure the patient's consent, fearing fireworks from the consultant if consent isn't given ("What do you mean the patient refused?").

Some patients are not as autonomous as you or me, one argument goes, so how can they truly give consent? Patients may be sick, frightened, embarrassed, or intimidated by the doctor. They may come from a culture where it is considered rude to question a doctor. They may wish to be a "good" patient—in other words, one who does not make a fuss. Even my friend with the missing submandibular gland, an unusually stubborn fellow, did not want to irritate the medical team. These emotional states are not conducive to autonomous choice; but is it not possible, by giving patients enough time, by creating a safe environment, by supporting patients and encouraging them to ask questions, to enhance a diminished autonomy sufficiently to get valid consent?

Time is the second highest barrier. Obtaining high quality consent usually takes more time than obtaining mediocre consent. I do not have a solution to the time problem, but it is worth noting how easy it is to use "lack of time" as a trump card against a tedious or unpleasant task. "Hurry is the devil," wrote William Osler, and in my biased view rushing consent should be avoided in the same way that a medical procedure should not be rushed. Both are bad medicine.

So what is the most redoubtable obstacle to valid consent? It is the still prevalent attitude that obtaining consent is a necessary chore, a medicolegal hurdle to jump over. Too often "consenting" a patient is reduced to the mechanistic imparting of information from clinician to patient or, worse still, the mere signing of a consent form, rather than the two way, meaningful conversation between clinician and patient it should be. If we can change this mindset and view obtaining consent as an ethical duty first and foremost, one that is central to respecting the autonomy and dignity of patients, then we will have taken a major step towards first class consent and uninterrupted lunches.

5.9   Appendix 9: What to Tell Junior Doctors About Ethics

Tomorrow I must give a talk to junior doctors. The title was imposed on me: "Essential ethics and law for the junior doctor." This may be the only hour they have on the subject in the entire year. What should be included in that hour?[9]

Consent is an obvious, unexciting choice. It is still the case that some junior doctors are asked to obtain consent for unfamiliar procedures; and, although some politely decline to do so, others do not want to make a fuss and acquiesce. And what of the patient who simply says, "I don't want to know—just do what's best, doctor"? Heaven also knows that some surgical patients are "consented" on the morning of the operation and have little idea of what awaits them. On the trolley heading for the operating theatre, one patient at a colleague's hospital told the porter that she was relieved at finally having the operation as she was looking forward to starting a family. She was about to undergo a hysterectomy. The porter called the medical team, and the operation was postponed.

---

[9] From Sokol D (2010) What to tell junior doctors about ethics, British Medical Journal, 340:c2489.

(If I have learnt one thing as a superannuated student and lecturer, it is that an ethics presentation without stories is like an operation without anaesthetic.)

Another "essential" issue is confidentiality. I shall not bore the junior doctors with old sayings about soundproof curtains and indiscreet discussions in the cafeteria. Instead I will focus on trickier scenarios, such as when to share confidential medical details with a patient's "partner" or when to breach confidentiality. The story this time will be of the patient who dies from a ruptured cerebral aneurysm during overzealous intercourse with his mistress. The distraught wife asks the medical team what happened. Discuss.

End-of-life decisions are another possibility, and there is much to be said about "do not attempt resuscitation" (DNAR) orders (*BMJ* 2009;338:b1723), quality of life, and the chameleon concept of "futility" (see Appendix 4), but junior doctors are unlikely to make such decisions in the near future. Still, they may be unsure about the exact implications for patient management of a DNAR order. Should they, for instance, start intravenous antibiotics on a DNAR patient? To close this section, a well placed anecdote concerning a grossly overtreated patient with cancer and the consultant's immortal words on reviewing the long list of procedures she had endured ("Jeez, it's hard to die in this hospital!") may stir them from their slumber and trigger a conversation on goals of care and the purpose of medicine.

This could lead to a discussion on the meaning of best interests. When we say that something is in the best interests of the patient, what do we mean? Examples from less conventional areas of medicine can provide a broader view of the concept. Sports doctors sometimes face a tension between clinical best interests and overall best interests, as when the patient, a professional boxer with a broken rib, wants to finish the round in the most important boxing match of his career. A prison doctor may also face a dilemma when she knows that a patient requesting diazepam is being coerced by some rough types to ask for the drug. The patient will not be treated kindly by the requestors if he fails to get some.

One option would be to talk more broadly about organisational ethics and problems with locums, rotas, continuity of care, targets, and patient safety. If I opened up the discussion, I could expect a torrent of stories about certain incompetent locums and, in the words of one of Eddie Murphy's film characters, locums who "don't speak English good" (It may be politically incorrect to say so, but safe and effective communication within the medical team and between patient and clinician is difficult without linguistic proficiency.) There might also be stories about government targets so slavishly followed that care of patients is undermined, and other dubious practices. NHS trusts, as public bodies, also have duties of care; and they can be sued for failing to provide adequate supervision or competent staff. This might be a good place to outline some law on clinical negligence, briefly looking at the standard of care and the controversial Bolam test, breach of duty, and causation. I will tell them that, after the 1988 case of Wilsher Vs Essex Area Health Authority, inexperience is not an excuse for negligent care and that calling your senior when unsure is legally, as well as medically, a very wise move.

Whistleblowing remains a problem, despite the Public Interest Disclosure Act 1998 and whistleblowing procedures adopted by NHS trusts. The story of junior doctors reporting their concerns about an underperforming colleague to a senior doctor only to see their concerns ignored, sometimes repeatedly and disdainfully, is a familiar one. The irony is that "the incompetent colleague" is a common question in membership examinations and job interviews, and candidates doubtlessly all give the right answer ("The care of my patient is my first concern"). The gap between the ideal world of General Medical Council guidelines and the clinical front line is a topic in itself.

Days could be spent on each of these issues, and I have ignored countless others, but the purpose of the session is not to provide the junior doctors with a solution to their problems (although I intend to give some answers at least) but to whet their ethical curiosity and provide them with a deeper appreciation of the pervasiveness of medical ethics. Most importantly, I would like them to leave the session with a spring in their step. What other profession can boast such a fascinating range of challenges and opportunities? That, perhaps, should be the essential message.

References

Charlton J (1990) The military quotation book. St Martin's Press

Kondro W (2007) Malaise in Marwais. CMAJ 177

Patterson K (2007) Talk to me like my father: frontline medicine in Afghanistan. *Mother Jones*. http://motherjones.com/politics/2007/06/talk-me-my-father-frontline-medicine-afghanistan?page=2

US Department of Defence (2004) Emergency war surgery. 3rd US revision, chapter 3

Sokol D (2006) Time to get streetwise. Br Med J 333:1226

Burns J, Truog R (2007) Futility: a concept in evolution. Chest 132:1987–93

Davis J Futility, conscientious refusal, and who gets to decide. Journal of Medical Philosophy 33:356–73

Fins J (2006) A palliative ethic of care. Jones and Bartlett, London

Gillon R (1997) "Futility"—too ambiguous and pejorative a term? J Med Ethics 23:339

Grossman E, Angelos P (2009) Futility: what cool hand Luke can teach the surgical community. World J Surg 33(7):1338–40

Jonsen A, Siegler M, Winslade W (2006) Clinical ethics, 6th edn. McGraw-Hill, New York

Kite S, Wilkinson S (2002) Beyond futility: to what extent is the concept of futility useful in clinical decision-making about CPR? Lancet Oncology 3:638–42

Russell I (2000) Consoler toujours—to comfort always. Journal of Musculoskeletal Pain 8:1

Schneiderman L, Jecker N, Jonsen A (1990) Medical futility: its meaning and ethical implications. Ann Intern Med 112:949–54

Sokol D (2008a) Clarifying best interests. Br Med J 337:a994

Dupuy J (1850) Considérations Pratiques Pour l'Amputation des Membres. Rignoux, Paris, France

Inman T (1860) Foundation for a New Theory and Practice of Medicine. John Churchill, London, UK

Kirkup J (2007) A History of Limb Amputation. Springer, London, UK

Porter R (1997) The Greatest Benefit to Mankind. HarperCollins, London, UK

Wootton D (2006) Bad Medicine: Doctors Doing Harm Since Hippocrates. Oxford University Press, Oxford, UK

Lichterman B (2008) Fyodor Grigorievich Uglov. Br Med J 337:a866

Osler W (1906) The fixed period. Aequanimitas, with other addresses to medical students, nurses and practitioners of medicine. McGraw-Hill, New York, pp 349–71

Starling P (2003) The case of Edward Revere Osler. Journal of the Royal Army Med Corps 149:27

General Medical Council (2006) Good medical practice. London. www.gmc-uk.org/guidance/good_medical_practice/duties_of_a_doctor.asp. Last accessed 31 July 2011

Sokol D (2009) When can doctors stay away? Br Med J 338:b165

Wynia M, Cummins D, VanGeest J, Wilson I (2000) Physician manipulation of reimbursement rules for patients. J Am Med Assoc 283:1858–65

Jones M (1999) Informed consent and other fairy stories. Medical Law Review 7:103–34

Manson N, O'Neill O (2007) Rethinking informed consent. Cambridge University Press, Cambridge

Sokol DK (2008b) Medicine as performance: what can magicians teach doctors? J R Soc Med 101:443–446